SPECIAL PROBLEMS
IN FIRE PROTECTION
ENGINEERING

Edited by

Paul R. DeCicco, P.E.

Volume IV
Applied Fire Science in Transition Series
Paul R. DeCicco: Series Editor

Routledge
Taylor & Francis Group

LONDON AND NEW YORK

First published 2002 by Baywood Publishing Company, Inc.

2 Park Square, Milton Park, Abingdon, Oxfordshire OX14 4RN
52 Vanderbilt Avenue, New York, NY 10017

Routledge is an imprint of the Taylor & Francis Group, an informa business

First issued in paperback 2019

Library of Congress Catalog Number: 00-041431

Library of Congress Cataloging-in-Publication Data

Special problems in fire protection engineering / edited by Paul R. DeCicco.
    p. cm. - - (Applied fire science in transition series ; v. 4)
  Includes bibliographical references and index.
  ISBN 0-89503-223-6 (paper)
    1. Fire protection engineering. 2. Fire prevention. 3. Fire extinction. I. DeCicco, Paul R., 1924- II. Series.

TH9210 .S66 2000
628.9'22- -dc21   00-041431

ISBN 13: 978-0-415-78491-7 (hbk)
ISBN 13: 978-0-89503-223-2 (pbk)

Contents

INTRODUCTION

Special Problems in Fire Protection Engineering

Building fires, because of the number of such structures that exist, the vast size of their aggregated occupancy populations, and the frequency and consequences of fires which occur in them, properly receive first attention from the fire safety community. However, all man-made structures are vulnerable to fire and the uniqueness of each particular (non-building) environment, and the fire hazards associated with them, make their protection the most challenging that fire safety professionals encounter. These structures are also significant in terms of their monetary value and because the potential for loss of life from fire may be extreme.

In this volume, the chapter by A. Michael Birk discusses engulfing fires in tanker trucks. The author uses TANKCAR to model the effects of fusible plug type temperature sensitive pressure relief valves and presents simulation results for different numbers of fusible plugs and for a fixed number of plugs to different melting temperatures. The temperature pressure relief valves (TPRV) simulation results are compared with experimental data and simulation results for an uninsulated tank filled with propane exposed to fire.

In their chapter on practical solutions for new fire protection problems, Kerwin and Forsythe discuss a general engineering approach to the modification of a bus facility to accommodate vehicles fueled by natural gas. Their second chapter, *Background on Facilities Modification for Natural Gas Fueled Bus Use*, reviews the analysis and requirements proposed for a typical natural gas bus facility. As a follow-up consideration, Sforza and Fox discuss the characteristics of compressed natural gas (CNG) and flammable plumes that may result from its leakage into the surrounding environment. The authors also propose an experimental setup to verify future theoretical analysis of gas leaks and to provide a source of data for those using CNG.

Smith and Kashiwagi in their chapter on *Expert Systems Applied to Spacecraft Fire Safety,* discuss the development of an expert system to manage fire safety in aircraft and in particular, NASA Space Station "Freedom." The authors present the unique fire safety problems and strategies for dealing with them. The chapter is of special interest in light of what has been learned about fires in low gravity environments and a number of fire events which have occurred on spacecraft since its writing.

Fernandez, Jacobs, Kauffman, Keating, and Sizemore discuss a decision analysis approach to assessment of community fire risk. The study was performed for accreditation under the National Fire Service Accreditation Program developed by the International Association of Fire Chiefs. The results of the study provide a model and risk assessment tool for fire science practitioners.

Ali and O'Connor present a case study on the calculation of axial forces generated in restrained pin-ended steel columns subjected to high temperatures. This chapter investigates the effect of the shortening of column height as the axially loaded pin ended element starts to lose its stability and deforms laterally.

Corbett's chapter on fire protection in large stadia (*The Alamodome in San Antonio*) discusses fire protection features incorporated in the design of the structure. Shortcomings of existing codes and standards necessitated the use of design fires to examine how the stadium and its occupants might respond to a particular fire scenario. The chapter also describes the fire testing of suppression and smoke management systems.

A systemic approach to fire safety in an offshore platform is presented in the chapter by Reyes, Beard, and Clark. The authors critique prescriptive and goal-setting approaches in addressing fire risks in this type of structure and describe a systems management procedure for analyzing fire safety issues.

Zhang, Fan, Shields, and Silcock, in two separate chapters, describe salt water techniques for simulation of fire smoke movement in an atrium building and smoke and induced air movement in a room-corridor building. In the atrium study, buoyant plume and circulation patterns were observed and circulation mechanisms were analyzed. In the room-corridor study, the occurrence of stratified smoke and air layers in the corridor were verified. The effects of position and heat release rate of the fire source on the smoke layer and the induced air layer are discussed.

CHAPTER 1

Modelling the Effects of Temperature Sensitive Pressure Relief Devices on Tankers Exposed to Engulfing Fires

A. Michael Birk

The present research involved the use of the Tank-Car Thermal Computer Model (TANKCAR) [1] to model the effects of fusible plug type temperature sensitive pressure relief valves (TPRVs) on an uninsulated tank. TANKCAR is a digital computer-based model that can simulate the thermal response of a long cylindrical tank filled partially with liquid and partially with vapor exposed to either an engulfing type fire, such as caused by a burning pool, or a torch type fire, such as that caused by a relief valve flare from a neighboring tank. The model can account for the effects of a number of thermal protection devices such as pressure relief valves, and novel internal protection devices including heat dissipating matrices. The model can simulate the effects of roll and pitch of the tank. Simulation results are presented for cases involving different numbers of fusible plug type TPRVs, and cases involving a fixed number of fusible plugs of different melting temperatures. In all cases the tanks are uninsulated with standard sized pressure relief valves and are filled with propane. The simulations involve engulfing type fires. The simulation results indicate that TPRVs could be very effective at reducing the risk of thermal ruptures of tanks. The TPRV simulation results are compared with experimental data and simulation results for an uninsulated tank filled with propane exposed to an engulfing fire.

Rail and highway tankers are equipped with thermal protection systems to minimize the effects of external heating such as may be caused by accidental fire impingement. Typical thermal protection systems include pressure activated pressure relief values (PPRV) and thermal insulation [2]. The PPRV's maintain the tank internal pressure below some safe value, and the thermal insulation material acts as a barrier to heat addition. The thermal insulation has the dual effect of

3

acts as a barrier to heat addition. The thermal insulation has the dual effect of maintaining wall temperatures below some critical value, and reducing the vapor generation rate within the tank [2].

The PPRV/thermal insulation combination has proven to be very effective. However, if for some reason the thermal insulation layer is damaged, and some of the tank wall is exposed directly to the effects of fire impingement, the wall temperatures in that region can increase dramatically. The resulting increase in wall temperatures can result in significant degradation of the tank wall material properties, and as a consequence the pressure carrying capability of the tank can be reduced. If the wall temperatures are allowed to go high enough, a thermal rupture of the tank will occur even if the PPRV is operating properly [2] to maintain the tank pressure below the nominal burst pressure of the tank.

One possible solution to this hypothetical situation is the temperature sensitive pressure relief valve (TPRV). This device acts to further reduce the tank internal pressure if high wall temperatures are sensed. Such a device would be sized to reduce the tank internal pressure well below that possible by the PPRV.

The TPRV can take different forms. Simple TPRV's can be valves with temperature sensitive elements that result in the opening of an orifice when the temperature increases. In the case of fusible plug type devices, a low melting temperature plug melts and is blown clear to render an orifice for pressure relief. More complex devices can be envisaged which involve data collection from strategically placed thermocouples, and the appropriate mechanisms to open a valve if temperatures exceed some critical value.

The present research involved computer simulations of tanks equipped with TPRV's filled with propane, exposed to engulfing type fires. The simulations were carried out using the tank-car thermal analysis program TANKCAR with modifications so that it could account for the effects of the TPRV's.

THE MODEL

The model used in the present research is based on the computer program TANKCAR [1]. This computer model can simulate a long cylindrical tank filled partially with liquid and partially with vapor exposed to either an engulfing type fire, such as caused by a burning pool, or a torch type fire, such as that caused by a relief valve flare from a neighboring tank. The model can account for the effects of a number of thermal protection devices such as pressure relief valves, thermal insulation, radiation shielding, temperature sensing relief valves, and novel internal protection devices, including heat dissipating matrices. The model is basically a two-dimensional representation of a circular cylindrical tank (i.e., axial gradients and end effects are not accounted for). However, the model has a pseudo-3D operating mode so that pitched and rolled tanks can be analyzed.

TANKCAR is capable of predicting the tank internal pressure, mean lading temperatures, wall temperature distribution, relief valve flow rates, liquid level, tank wall stresses and tank failure all as functions of time from initiation of the fire impingement. These various outputs define the response of the tank/lading system and provide valuable information for the design of a thermal protection system.

The model has been extensively validated [3] by comparing its predictions with the results of numerous fire tests involving full [4] and fifth-scale [5,6] rail tank-cars exposed to engulfing fires.

TANKCAR is made up of a series of submodels simulating the following processes:

1. flame to tank heat transfer,
2. heat transfer through the tank wall and associated coverings,
3. interior-surface to lading heat transfer,
4. thermodynamic process within the tank,
5. thermodynamic properties of the lading,
6. pressure relief device operating characteristics,
7. wall stresses and material property degradation, and
8. tank failure.

The flame-to-tank heat transfer submodels can account for either an engulfing pool fire or a two-dimensional torch. The pool fire model accounts for both thermal radiation and free convection heat transfer from the fire to the tank. The thermal radiation is calculated as a function of the circumferential position on the tank surface by accounting for the typical shape of large pool fires in the absence of cross wind effects. The convection calculations are based on empirical relations for convection to horizontal cylinders in a crossflow. The torch submodel is based on empirical relations for the heat transfer from a jet impinging on a flat plate. The geometry of the jet/plate system in the submodel is similar to the U.S. DOT Transportation Test Center Torch Simulator as described in Reference [7].

The heat transfer through the tank wall and associated coverings is represented using finite difference techniques. The finite difference solution accounts for pure conduction through the tank wall and insulating layers, and pure conduction, convection, and radiation when appropriate, such as in the case of certain heat conducting matrices. The finite difference solution also accounts for the heat transfer from the fire to the tank outer surface and from the tank inner surface to the lading. The interior surface heat transfer submodels account for convection and radiation in the vapor space, and convection and boiling in the liquid region. The radiation calculations are based on the assumption that the wall communicates only with the lading. The convection and boiling coefficients are based on empirical relations for inclined flat surfaces.

The thermodynamic process submodel treats the lading as three distinct regions—the vapor space, the liquid boundary, and the liquid core. It is assumed that the vapor and liquid boundary (near wall, and free surface) are in thermodynamic equilibrium and saturated; the core is assumed to be subcooled initially, but after some period of venting it is assumed to be in equilibrium with the liquid boundary and vapor space. This submodel requires the setting of two empirical constants, the liquid boundary thickness, and the energy partition factor that determines how much of the fire heat is transferred into the vapor and liquid boundary, and how much is transferred to the liquid core. These constants have been determined by calibrating the model with one set of fire test results (full-scale tank fire test [5]) and have remained unchanged for all subsequent validation runs and simulations. In the submodel it is assumed that the energy for venting comes from the vapor and liquid boundary. This energy drain is one reason that the liquid core eventually reaches equilibrium with the other regions.

The thermodynamic and transport properties of the lading are based on the Starling equation of state [8], and on available material property data, respectively. The pressure relief valve submodel accounts for both the mechanical action of a relief valve and the fluid mechanics. The valve mechanics are accounted for using a steady state model. Valve cycling dynamics are not accounted for explicitly, but rather implicitly by having a model valve that remains partially open to represent the reduced flow capacity of a real valve during cycling, and fully open when in reality the valve is fully open. The opening fraction is determined by the lading vaporization rate. The valve flow submodels can account for choked dry vapor flow and for frozen liquid flow.

The wall stresses are calculated at the point on the wall circumference which experiences the highest temperature and includes pressure induced (hoop) stress and stresses due to radial temperature gradients in the tank wall. The tank failure analysis is based on the maximum normal stress theory of failure. Degradation in the wall material strength with increases in temperature is accounted for using available data for tank-car steel.

Further technical details about these various submodels are given in Reference [1].

TPRVs

The TPRVs were modelled as simple fusible plug devices that were either closed or open depending on the temperature of the tank wall where the devices were located. When the temperature of the wall at the location of the plug reached a preselected temperature it was assumed that the plug melted and was blown clear thereby providing an orifice through which the tank lading could be vented. Since it is unlikely that practical melting temperatures will be reached in wall areas below the lading liquid level, it was assumed that only vapor would be flowing through the orifice.

When the plugs melted out it was assumed that they remained open even when the wall temperature where they were located dropped below the melting temperature of the plug. In other words the devices were modeled as being irreversible. This is not representative of all TPRVs but it is reasonable for fusible plug type devices.

When the orifice is open in the TPRV the vapor flow is choked for the tank pressures of interest. The equation used to calculate mass flows through the orifices was that for isentropic one dimensional flow in varying area channels [9]. A coefficient of discharge of 0.8 was used for all fusible plug flow calculations.

VALIDATION

The overall tank-car model has been validated [3] using available experimental data from full and fifth-scale fire tests of rail tank-cars. An example of validation results is presented in Figures 1 to 4 which show TANKCAR predictions and experimental results for a full-scale uninsulated tank-car with a standard sized relief valve (DOT class 112/114A340) exposed to an engulfing fire. The tank was filled to 95 percent capacity with propane. The experimental data points were obtained from Reference [4]. As can be seen from the Figures, the model predictions are in good agreement with the results of the experiments. For additional examples of overall model validation the reader is directed to Reference [3].

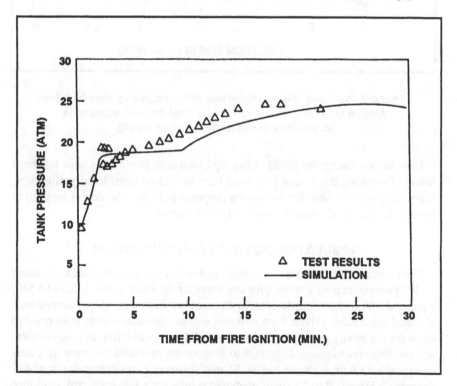

Figure 1. Predicted and measured tank pressure vs. time from fire ignition for a full-scale upright uninsulated tank-car exposed to an engulfing fire (data from Reference [4]).

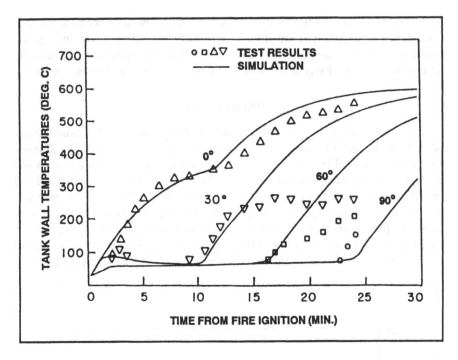

Figure 2. Predicted and measured wall temperatures vs. time from fire
ignition for a full-scale upright uninsulated tank-car exposed to
an engulfing fire (data from Reference [4]).

Data for validating the fusible plug equipped tank predictions have not been
found. Therefore, the results presented here should be considered preliminary.
They are presented here for discussion purposes only and should not be used to
govern the design of actual thermal protection systems.

SIMULATION RESULTS AND DISCUSSION

The simulations about to be presented involve full-scale uninsulated tanks filled
to 95 percent capacity with propane (similar to DOT class 112/114A340)
equipped with fusible plug devices. In all cases the tanks were also equipped with
standard sized relief valves. Two different sets of simulations were performed to
show the sensitivity in the performance of fusible plugs to different plug numbers
and melting temperatures. Figures 5 to 8 show the predicted response of a tank
equipped with four, eight, or twelve 51 mm diameter plugs that melt out at 425
degrees C. Figures 9 to 12 show predicted results for a full-scale tank equipped
with four 51 mm diameter plugs that melt out at temperatures of approximately
205, 315 and 425 degrees C.

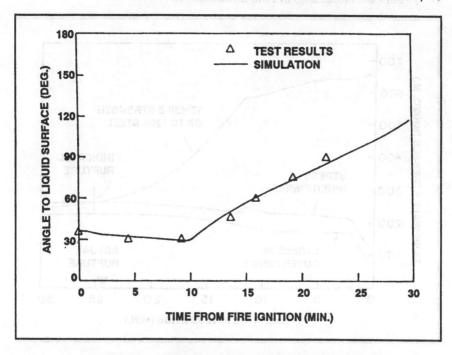

Figure 3. Predicted and measured liquid level vs. time from fire ignition for a full-scale upright uninsulated tank-car exposed to an engulfing fire (data from Reference [4]).

In all the simulations, the plugs were located along the tank top-dead-center (tdc). This, however, should not be regarded as a recommendation that fusible plugs be located only at the tank tdc. Fusible plug devices should be located strategically over the tank surface to ensure that at least some remain above the liquid surface for all possible tank orientations. This is necessary because the fusible plugs are unlikely to melt out as long as they remain liquid wetted (i.e., wall areas which are below the liquid level are effectively cooled by the liquid, and therefore wall temperatures in areas that are liquid wetted remain close to the liquid temperatures). The optimum placement of fusible plug devices requires further study which is beyond the scope of the present project.

For comparison purposes the reader should refer back to Figures 1 to 4 which show simulation and test data for an unprotected full-scale tank exposed to engulfing fire conditions. In the following discussion reference will be made to these Figures. Those interested in comparing the present results with similar predictions for insulated tanks are directed to Reference [3].

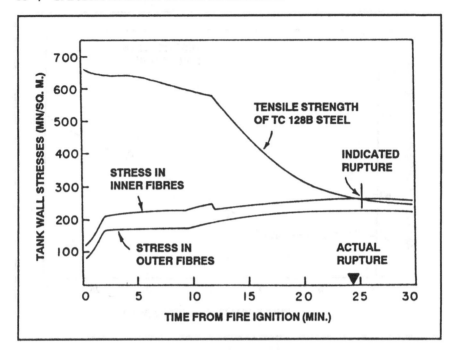

Figure 4. Predicted tank wall stresses and material strength vs. time
from fire ignition for a full-scale upright uninsulated tank-car exposed to an
engulfing fire (data from Reference [4]).

Case #1: Different Number of Fusible Plugs

Figure 5 presents the predicted tank pressure for tanks with different numbers of fusible plugs. As can be clearly seen from the Figure, the tank response prior to the opening of the fusible plugs is identical to that for an unprotected tank-car. However, once the fusible plugs come into play, the tank pressure histories differ significantly from that for the unprotected tank. As is indicated in Figure 5, increasing the number of fusible plugs increases the rate at which the tank pressure drops after the plugs open. The law of diminishing returns appears to apply with the number of fusible plugs. It seems that there is not a great advantage to having more than eight fusible plugs in this specific example.

Figure 6 presents the predicted tank wall temperatures for the tanks with different numbers of fusible plug devices. As can be seen from this Figure, the presence of fusible plugs has no effect up until the point where the fusible plugs melt. This was also seen in the previous Figure that showed the predicted tank pressure. Once the fusible plugs melt and material is vented through the devices, the wall temperatures are affected. This is because the lading is vented more rapidly to the atmosphere in the fusible plug equipped tank than in the unprotected

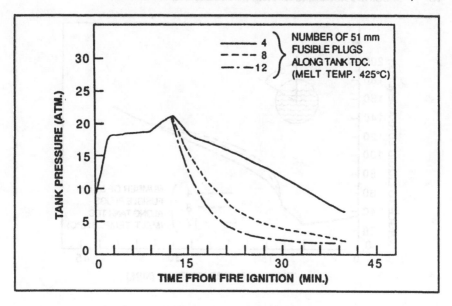

Figure 5. Predicted tank pressure vs. time from fire ignition for full-scale upright uninsulated tank-cars, equipped with different numbers of fusible plug devices, exposed to engulfing fires.

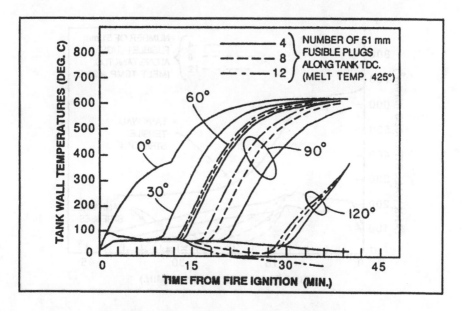

Figure 6. Predicted wall temperatures vs. time from fire ignition for full-scale upright uninsulated tank-cars, equipped with different numbers of fusible plug devices, exposed to engulfing fires.

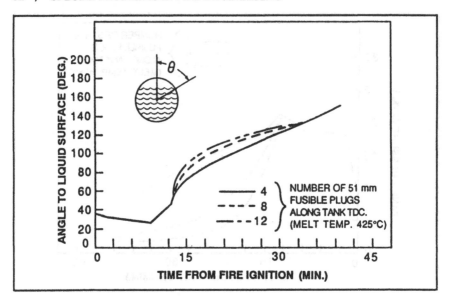

Figure 7. Predicted liquid level vs. time from fire ignition for full-scale upright uninsulated tank-cars, equipped with different numbers of fusible plug devices, exposed to engulfing fires.

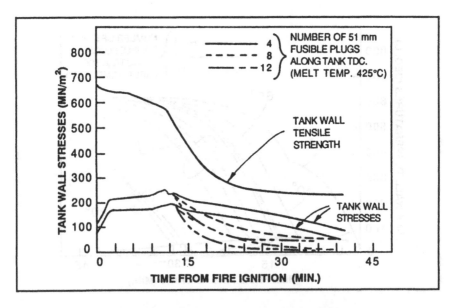

Figure 8. Predicted tank wall stresses and material strength vs. time from fire ignition for full-scale upright uninsulated tank-cars, equipped with different numers of fusible plug devices, exposed to engulfing fires.

tank. As the liquid level drops in the tank, more of the tank wall is exposed to vapor and as a result the wall temperatures in these exposed areas rise very rapidly. If the liquid level drops more rapidly in the fusible plug equipped tank as the number of plugs is increased then this will of course effect the wall temperatures.

Figure 7 presents the predicted liquid level for the different cases considered. As can be seen from the Figure, once the fusible plugs open there is a sudden increase in the rate at which the liquid levels drop. Initially the tank with the largest number of fusible plugs experiences the most rapid venting rate because this case provides the largest area for relief flow. However, the liquid levels for the different cases eventually converge to nearly the same values. This can be explained as follows. As the tank pressure drops the rate at which lading is vented decreases. Therefore, by increasing the number of fusible plugs, one decreases the pressure at the same time one increases the total orifice area. This results in the two cases giving similar venting rates. Also, it must be remembered that the venting rates are ultimately limited by the total heat input to the tank and this total heat flux is very similar for the different cases considered.

Figure 8 presents the predicted tank wall stresses and tank wall material strength for the different cases considered. The fusible plugs have little effect on the material strength of the tank wall when compared to the case of the unprotected tank. This is because the peak wall temperatures are unaffected by the presence of the fusible plugs, and it is these temperatures that determine the weakest point in the tank wall. The other two curves in the Figure show the tank wall stresses at the inner and outer fibers of the tank wall. As can be clearly seen from the Figure, the tank wall stresses are significantly reduced from those in an unprotected tank (see Figure 4) when the fusible plugs are opened. This is because the tank pressure has been reduced and this results in a proportional decrease in the tank hoop stresses. Once again, the law of diminishing returns is clearly indicated in the predicted results. There seems to be little advantage in increasing the number of fusible plugs above either in this specific example.

In summary, it appears that there is an optimum number of fusible plugs to protect a tank-car. The optimum number of plugs will be determined by a balance of economic and thermal considerations. It appears that four fusible plugs, each of a diameter of 51 mm melting out at 425 degrees C, is an effective means of protecting a tank-car from an engulfing fire environment. This assumes, however, the four plugs are all above the liquid surface (e.g., at the tank top-dead-center, and the tank is upright and horizontal) at the initiation of the fire. Plug placement has not been addressed here for cases where the tank is rolled or pitched.

Case #2: Fusible Plugs of Different Melting Temperatures

Figure 9 presents the predicted pressures for the tanks equipped with a fixed number of plugs of different melting temperatures. As in the previous case the

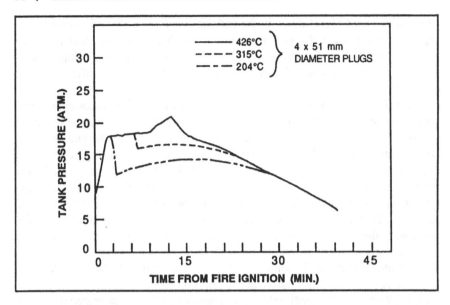

Figure 9. Predicted tank pressure vs. time from fire ignition for full-scale upright uninsulted tank-cars, equipped with different melting temperature fusible plug devices, exposed to engulfing fires.

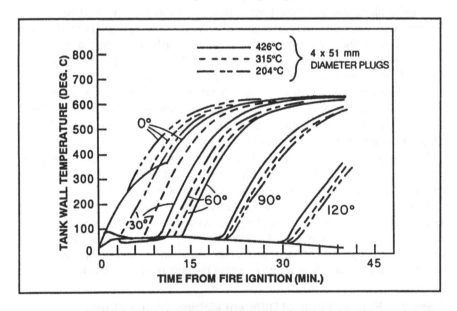

Figure 10. Predicted wall temperatures vs. time from fire ignition for full-scale upright uninsulated tank-cars, equipped with different melting temperature fusible plug devices, exposed to engulfing fires.

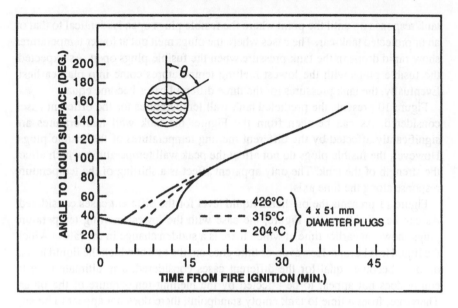

Figure 11. Predicted liquid level vs. time from fire ignition for full-scale upright uninsulated tank-cars, equipped with different melting temperature fusible plug devices, exposed to engulfing fires.

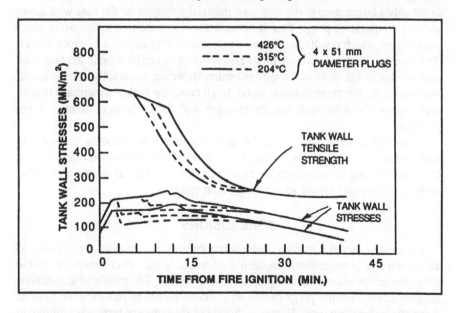

Figure 12. Predicted tank wall stresses and material strength vs. time from fire ignition for full-scale upright uninsulated tank-cars, equipped with different melting temperature fusible plug devices, exposed to engulfing fires.

tank response up until the point where the fusible plugs open is identical to that of an unprotected tank-car. The cases where the plugs melt out at lower temperatures show rapid drops in the tank pressure when the fusible plugs open. As expected, the fusible plugs with the lowest melting temperatures come into play earliest. Eventually, the tank pressures for the three different cases become equal.

Figure 10 presents the predicted tank wall temperatures for the different cases considered. As can be seen from the Figure, the tank wall temperatures are significantly affected by the different melting temperatures of the fusible plugs. However, the fusible plugs do not affect the peak wall temperatures which affect the strength of the tank. The only apparent affect is a shifting of the temperature response along the time axis.

Figure 11 presents the predicted liquid level for the different cases considered. As can be seen from the Figure, the case with the lowest melting temperature plugs shows an earlier time at which there is a sudden change in the rate at which the liquid level drops. However, as time goes on, the rate at which the liquid levels drop all become equal for the different cases considered. The ultimate time to empty does not appear to be affected by the melting temperature of the plugs. Therefore, from a time to tank empty standpoint, there does not appear to be any advantage in reducing the melting temperature of the plugs.

Figure 12 presents the predicted tank wall stresses and tank wall material strength for the different cases considered. As can be seen from the Figure, there is actually a larger drop in the tank wall material strengths for the case with lower melting temperature plugs. This is because the lower melting temperature plugs open earlier which results in the lower wall areas being exposed to vapor earlier. The lower melting temperature plugs also result in an earlier drop in the tank wall stresses due to the drop in the tank pressure. However, eventually, the tank wall stresses for all the cases become equal. In all cases the tank wall stresses remain well below the tank wall tensile strength and therefore tank failure is not predicted.

In summary, there appears to be little advantage in reducing the melting temperature of fusible plugs. In fact, there is actually a detrimental effect caused by lower melting temperatures because the risk of accidental or unwanted plug melting is increased for the lower temperature plugs.

CONCLUSIONS

A computer model capable of simulating the response of tank-car exposed to accidental fire impingement has been used to study the effectiveness of fusible plug type temperature sensitive pressure relief devices. The results of simulations suggest that the fusible plugs can be an effective means of reducing the risks of thermally induced ruptures. However, further study including testing is required to establish the true benefits of using such devices on actual tankers.

REFERENCES

1. A. M. Birk, *User Documentation for the Rail Tank-Car Thermal Analysis Program,* Report No. 83-29, Davis Engineering Ltd., Ottawa, March 1985.
2. W. Townsend, C. Anderson, J. Zook, and G. Cowgill, Comparison of Thermally Coated and Uninsulated Rail Tank Cars Filled with LPG Subjected to a Fire Environment, *Report NO. FRA-OR&D,* pp. 75-32, December 1974.
3. A. M. Birk, Validation and Testing of the Tank-Car Thermal Analysis Program, *Transport Canada Report NO. TP6693E,* Davis Engineering Ltd., Ottawa, March 1985.
4. Railroad Tank Car Safety Research and Test Project, *Phase II Report on Full-Scale Fire Tests,* AAR-RPI, RA-11-6-31, AAR-R-201, US6 AFI 17-75 ENG(BK).
5. Railroad Tank Car Safety Research and Test Project, *Phase II Report on Analysis of Fifth-Scale Fire Tests,* AAR-RPI, RA-11-5-26, December 1973.
6. R. D. Appleyard, Testing and Evaluation of the Exposafe System as a Method of Controlling the Boiling Liquid Expanding Vapour Explosion, *Transport Canada Report TP 2740,* August 1980.
7. C. E. Anderson, *Rail Tank-Car Safety by Fire Protection,* 6th International Fire Protection Seminar, 1982.
8. K. E. Starling, *Fluid Thermodynamic Properties for Light Petroleum Systems,* Gulf Publishing Company, 1973.
9. A. H. Shapiro, *The Dynamics and Thermodynamics of Compressible Fluid Flow,* Ronald Press, 1953.

CHAPTER 2

Practical Engineering for New Fire Protection Problems: An Application to Natural Gas Fueled Bus Facilities

Ralph Kerwin and Thomas J. Forsythe

As the published body of fire protection engineering research grows, the successful application of rigorously validated theoretical models increases. However, the practicing design professional still encounters a large gap between theory and practice when faced with a complex, novel fire protection problem. The first chapter in this two-part series illustrates the general approach used by a consulting fire protection engineering firm to develop a set of practical fire safety requirements for a situation which lacks any direct, pre-existing code guidance—the modification of a bus facility to accommodate vehicles fueled by natural gas. The second chapter reviews the analysis and the requirements proposed by the authors' consulting firm for a typical natural gas bus facility.

As the published body of fire protection engineering research grows, applications of validated theoretical models are increasing. However, a practicing design engineer is still often faced with a large gap between fire protection theory and practice when facing an unusual fire protection problem. A thorough and far reaching literature search is necessary in order to develop effective and practical requirements which have as strong a basis as possible in existing empirical knowledge and research findings.

The practicing fire protection engineer typically lacks the resources to do extensive original research or experimentation. In addition, adequate funding for comprehensive research to address a rapidly developing new fire protection challenge is frequently unavailable in the early stages of development of an engineering response to the hazard. To effectively assess a new hazard, the engineer must approach the problem in a manner analogous to the initial stages of a scientific experiment. Before conducting an experiment, the scientist first produces a

19

testable hypothesis and then conducts a literature search to find whether existing information supports or disproves the hypothesis. The results of this search may alter the original hypothesis. Based on the modified hypothesis, plans are made for a detailed experiment. For a design engineer, the end objective is a set of engineering recommendations for a client rather than a plan for an experiment. The method consists of the following four steps:

1. Create a hypothesis (hazard analysis) which divides the problem into components which can be analyzed in a meaningful fashion.
2. Conduct a broad search for existing literature and other information which can be applied to these components in an attempt to validate (or disprove) them. Frequently, findings must be adapted from research conducted in related fields and applied using considerable judgement.
3. Assess the quality and appropriateness of the gathered information. A substantial amount of physical reasoning and judgement may be required to verify that literature on analogous topics can be conservatively applied to the hazard under study. Modify the model as necessary based on the results of the assessed findings.
4. Provide recommendations to mitigate or alleviate the various components of the problem based on the modified hazard model.

This chapter provides an example of such an approach. The authors acknowledge sole responsibility for application of any cited research beyond its original scope of application.

Confronting an unusual fire protection problem as a consultant can be a daunting task. At least two missteps are possible. First, engineers under budgetary and schedule pressures may be tempted to apply inappropriate criteria which appear superficially similar to the hazard under study. As an example, National Electrical Code guidelines are sometimes cited out of context to justify electrical classification throughout a natural gas bus storage and maintenance building—an undertaking which is expensive, difficult to maintain and, in the authors' judgement, unnecessary. Second, an engineer may call for extensive experimental work to model the hazard without sufficient study and planning. While well planned research is of great value, a lack of planning can lead to costly research which merely replicates results which could have been obtained from basic analysis or the existing literature. In either case, these approaches poorly serve the intended user. Inappropriate engineering designs become especially important when the new hazard is due to the rapid adoption of a new technology. Given the hundreds of municipal transit authorities throughout the United States which are quickly acquiring natural gas fueled vehicles, the cumulative financial consequences of retrofitting bus facilities with inappropriate or unnecessary measures is enormous.

The ultimate intent of the design process is to provide practical fire protection measures which mitigate the fire protection hazard to a level acceptable to the design engineer, to the authorities having jurisdiction and to the client. To this end, the engineer takes the following steps.

PRELIMINARY HAZARD ANALYSIS

First, a hypothesis or hazard analysis must be created modeling the perceived effects of the hazard under study.

1. Conduct a preliminary analysis of the hazard based on readily available information. Qualitative measures of severity and probability should be assigned based on available information and the application of informed engineering judgment. In addition, the different aspects of the hazard should be characterized (for instance, what type of natural gas fuel system accidents would produce specific types of leaks). Qualitative risk assessment techniques from the field of safety engineering are quite valuable for this purpose [1-2].

2. Isolate the various aspects of the hazard into components. These components should allow for a meaningful examination of the hazard in a piecemeal fashion. For instance, dividing the possibility of natural gas leakage from bus fuel systems into "slow" and "fast" leaks and assessing their relative probabilities and severities allows for separate engineering solutions to be applied to each case. "Slow" leaks, even from multiple buses, have a much higher probability of occurrence yet can be addressed using readily achievable ventilation solutions. This narrows the area of concern to less probable "fast" gas releases. Anyone who has assembled a jigsaw puzzle is familiar with this process; one fits the easiest portions of the puzzle together first while isolating successively more difficult portions for later work.

INFORMATION RESEARCH

Second, all available sources of information should be utilized which may pertain to the hazard under study. The need for a comprehensive literature search is frequently overlooked by both experimenters [3] and designers.

1. Conduct a search of current literature in the field. In truly novel fire protection applications, the quantity of directly applicable literature is relatively meager.

2. Conduct extensive telephone interviews of key parties such as manufacturers, users and professional societies to learn current information not found in the available literature. The importance of developing knowledgeable "contacts" who can share topical information cannot be overemphasized. Attendance at relevant seminars and conferences can be particularly helpful in this regard.

3. Develop an understanding of the user's operations and concerns by making site visits to the user's facility and by reading topical literature from the user's field. Once again, attendance at conferences sponsored by the user's professional organizations is beneficial. For instance, bus transit authorities are represented by the American Public Transit Association. Understanding the user's terminology,

concerns and operations are important since recommendations which will substantially interfere with the user's day to day operations are likely to meet resistance, regardless of their intrinsic fire safety merit. Conversely, recommendations which may seem unfeasible to the engineer may be quite achievable when combined with technology familiar to the user. For example, a major bus facility pointed out that they were adapting a special radio system to aid in location of their buses inside their storage facility and to accommodate disabled passengers. They noted that it would be a simple step to add supervised, on-board gas detection to their natural gas buses that would be received by the facility management system in real time. This feature will substantially improve the timeliness and reliability of natural gas leakage detection in the facility and will allow them to avoid other costly facility modifications, such as extensive electrical classification near ceiling level [4].

4. Broaden the literature search to include other specialties which might conceivably have fire safety data relevant to the hazard. For instance, based on preliminary analysis, it was determined that the petrochemical industry would have a substantial body of literature relating to the handling and behavior of compressed and liquified flammable gases. Sometimes, the most pertinent literature is based on studies contained in journals published twenty or more years previously.

LITERATURE ASSESSMENT

Third, the engineer must determine precisely how the gathered information can be applied to support the preliminary hazard model. Care must be taken when extrapolating such information to ensure that it applies to the hazard in a conservative fashion. For instance, in an early assignment studying the applicability of electrical classification to natural gas fueled bus facilities, one of the authors conducted an extensive analysis to demonstrate that certain basic principles embodied in the use of electrical classification for chemical plants could be extended to apply to natural gas fueled bus garages [5]. The original intended application of the published research will not generally be identical to the hazard under study; it is sufficient to show that the results of interest conservatively bound one or more aspects of the hazard. For example, the findings of NPFA 30, Appendix F [6], are intended to apply to so-called "fugitive emissions" resulting from the handling of flammable liquids. However, the pertinent results from this analysis can be shown to be at least as effective in describing the handling of natural gas [7].

Although the process of designing against a low probability, high consequence fire protection hazard may not possess the degree of rigor of a scientific experiment, the engineer must continually assess uncertainty in the information and ask "How well supported are the various elements of the analysis?"

The literature assessment may support the preliminary hazard analysis or may prompt refinement of the analysis. Ideally, the various findings can be fit together to seamlessly support all aspects of the hazard analysis. In practical terms, this often does not happen and less rigorous forms of analysis, such as anecdotal evidence, are used to "fill in the cracks." The rigor (and thus, desirability) of supporting information is ranked as follows:

1. Multiple, independent and overlapping publications addressing both theory and experimentation can be extrapolated to support one or more components of the hazard analysis or design recommendations (for instance, the tendency of methane to be sharply buoyant at room temperature).

2. A single authoritative publication (e.g., an NFPA standard) can be extrapolated to support one or more components of the hazard analysis or design recommendations (for instance, the analysis of the applicability of NPFA 497A and 497M electrical classification recommendations to bus facilities).

3. Original analysis conducted by the Engineer, applying fundamental scientific or engineering principles to a straightforward circumstance (for instance, the use of thermodynamic principles to bound the lower temperatures of natural gas being evacuated rapidly from a high pressure cylinder).

4. One or more publications offering suggestive but not rigorous or conservative support. For example, an unvalidated finite element analysis demonstrates the behavior of natural gas following a catastrophic release in a bus facility [8]. An engineer must be particularly careful when relying on unvalidated computer modeling to establish critical results.

5. Anecdotal evidence (preferably but not necessarily published) used in conjunction with engineering judgement. Anecdotal evidence is particularly valuable for identifying failure modes in new technological applications. For example, anecdotal evidence identified a failure mode for a particular style of pressure relief device which would otherwise have been assumed to be of negligible probability. While studies have been done on the effects of corrosion on the interiors of compressed natural gas cylinders [9], anecdotal evidence also helped to identify failure modes due to physical or chemical damage to cylinder exteriors.

6. Engineering judgement used without corroboration (should be applied with strong restraint). Areas in which engineering judgement must be relied on for important assumptions are areas which should be called out for applied research or further study.

Although "engineering judgement" is used here specifically to indicate the postulation of hazard behavior without strong empirical or theoretical support, it must be emphasized that experienced judgement plays a crucial role throughout the conduct of any design.

RECOMMENDATIONS

Fourth, propose recommendations to control the analyzed hazard to a reasonable level based on an understanding of the limitations of various fire prevention or mitigation measures and the needs of the user. This is the heart of conceptual design. The desirability of recommendations is based on the following hierarchy:

1. Engineered "passive" safeguards and constantly active, interlocked safeguards (for example, designing bus fuel systems to prevent or discourage large scale leaks or providing continuous ventilation which will sound an alarm if it fails),
2. Engineered automatically activated safeguards (for instance, emergency power shutdown initiated by methane detection),
3. Operational procedures involving human action (for instance, hot work procedures and proper training for all natural gas fueled bus mechanics).

Protection of life and property are the primary issues in designing for fire prevention or mitigation. The following issues are strong secondary concerns:

- Potential impact on user operations
- Maintainability of safety equipment
- Cost of proposals relative to perceived property conservation benefits

IDENTIFYING AREAS FOR ORIGINAL RESEARCH

During the engineering design process, a number of topics are often noted which deserve further study. Typically, suggestions for such study would represent requests for applied research to answer specific *physical questions* which could not be fully addressed by a comprehensive literature search, or to justify judgements which lack strong theoretical or empirical validation. In practical terms, clients are frequently not interested in funding such research unless it is of critical importance to the design. This is rarely the case, as the design can often be modified so as to avoid sensitivity to the unresolved issues. For instance, even though the precise amount of time that a flammable gas cloud will linger at ceiling level under given ventilation conditions is unknown, a ceiling level zone of electrical classification can be arranged to deprive such a cloud of potential electrical ignition sources.

DISCUSSION

While the process given above appears to be linear, proceeding one step after the next, the development of a comprehensive design to address a new hazard tends to work as an iterative process—additional information causes reassessment or refinement of preliminary conclusions. Since, by definition, there are no experts on truly new fire protection problems, much of the development process begins as an open-minded outreach for information. Given the large and rapidly growing body of technical information, conducting such a search is more of an art than a science. Pertinent research may be decades old or "hidden" in fields of study unfamiliar to the engineer. The essential creative act for an engineer in this process is to seek out useful information. Key qualities worth cultivating for such an endeavor are a curiosity regarding related technical areas and the willingness to grapple with new concepts and fields of study and relate their significance to the hazard under review. The engineer must recognize and apply potential similarities in fundamental principles between apparently disparate situations.

Some practical issues are worthy of mention:

1. A thorough literature search will result in the accumulation of substantial quantities of written material. Proper organization of these materials allows for easy retrieval.

2. A surprisingly large quantity of valuable hints regarding pertinent literature is found not by any innate talent of the Engineer but rather by virtue of establishing numerous contacts within the affected industries. Frequently, manufacturers, users, and other interested parties will share information or literature which seems to them "routine" but which is valuable to the Engineer. A generous approach to sharing information and a courteous telephone manner are of great assistance. Think over potential questions in advance in order to draw out the most useful information.

3. The value of access to technical resources cannot be overstated. Sometimes, the most helpful information is "stumbled upon" while browsing, or speaking with other experienced engineers. Resources of value include engineering data bases such as NIST's FIREDOC, and membership in technical organizations such as the Society of Fire Protection Engineers (SFPE).

4. A significant quantity of time is necessary for all of the activities mentioned above. Cultivating industry contacts and pursuing promising leads over the phone is rewarding but very time consuming. The authors' consulting firm has been strongly involved in the issue of natural gas fueled buses for over two years as of the writing of this article, admittedly balanced with other ongoing engineering pursuits. The fundamental concepts of compressed natural gas (CNG) facility fire protection developed over a period of approximately two months, but the full evolution of CNG recommendations involved a part time effort involving two people over 18 months and several projects. Due to the vicissitudes of business development, detailed development of liquified natural gas (LNG) recommendations began substantially later than the CNG recommendations. The refinement of LNG recommendations is ongoing.

SUMMARY

The process of designing against a new type of fire hazard resembles the planning of an experiment. In both cases, a hypothesis (hazard analysis) must be created and refined, literature and other information must be reviewed and assessed, and recommendations must be made. The next chapter of this two-part series will examine the hazard analysis and recommendations for a hypothetical natural gas fueled bus maintenance and storage facility.

REFERENCES

1. MIL-STD-882C, *Military Program Standard for System Safety Program Requirements*, Department of Defense, January 1993.
2. J. W. Vincoli, *Basic Guide to System Safety*, Van Nostrand Reinhold, New York, 1993.
3. J. P. Holman, *Experimental Methods for Engineers*, McGraw Hill, New York, pp. 29-31, 1984.
4. R. Kerwin, *Fire Safety Implications of Conversion to Accommodate CNG-Fueled Bus Use at the Mitchel Field Bus Complex*, Draft Report, Gage-Babcock & Associates, pp. 3-10, August 1994.
5. R. Kerwin, *Case Study: Electrical Classification for a Natural Gas Vehicle Maintenance Facility*, National Fire Protection Association Presentation, April 1993.
6. NFPA Standard 30, *Flammable and Combustible Liquids Code*, National Fire Protection Association, pp. 68-69, 1993 Edition.
7. R. Kerwin, *Fire Safety Implications of Conversion to Accommodate CNG-Fueled Bus at the Mitchel Field Bus Complex, Appendix C*, Draft Report, Gage-Babcock & Associates, pp. A-4-A-6, August 1994.
8. M. Murphy, S. Brown, and D. Philips, *Extent of Indoor Flammable Plumes Resulting from CNG Bus Fuel System Leaks*, Battelle, SAE Technical Paper 922486, November 1992.
9. F. Lyle, Jr. and H. Burghard, Jr., *Effects of Natural Gas Contaminants on Corrosion in Compressed Natural Gas Storage Cylinders*, Southwest Research Institute, SAE Technical Paper 861544, 1986.

CHAPTER 3

Background on Facilities Modification for Natural Gas Fueled Bus Use

Thomas J. Forsythe and Ralph Kerwin

In recent decades, the fire safety concerns of vehicle maintenance and storage garages have not been a strong focus in the fire protection literature. The long-standing preeminence of diesel oil and gasoline as vehicle fuels has led to widely accepted fire prevention measures present in all major fire codes. These traditional approaches to fire safety in vehicle maintenance and storage facilities are being questioned, however, as federal and state mandates create a preference for vehicles which run on natural gas instead of diesel oil or gasoline. These questions have been raised in a particularly urgent fashion among the nation's mass transit bus systems. For several reasons, bus fleets have been singled out by the federal government for a particularly rapid transition to alternate fuels. Many transit agencies are opting for natural gas in either compressed or liquefied (cryogenic) form. Facilities managers across the country face the same question, "How do I handle fire safety concerns when I retrofit my vehicle storage and maintenance facility for natural gas?" Currently, no codes or standards directly address this issue.

To demonstrate one possible approach to this problem, the following sections of this chapter will describe a hypothetical facility layout, a description of a typical compressed natural gas (CNG) fuel system, a hazard analysis and accompanying recommendations. This hypothetical scenario is based on typical features of several actual studies accomplished by the authors. A similar treatment is then accorded liquefied natural gas (LNG).

27

DESCRIPTION OF BUS FACILITY

This chapter will review a generic garage setting, similar in concept to many located in colder climates (Figure 1). It differs from facilities in warmer climates in that overnight bus storage and bus servicing are conducted indoors. This simplified facility has areas for maintenance (typical of all transit authorities), areas for servicing, and areas for indoor bus storage (usually found only in colder climates). Administrative office areas adjoin the garage. The ceiling is slightly over 20 feet high and is supported by an open truss arrangement that allows unobstructed air flow near ceiling level. The facility has been using diesel fueled buses and is now considering a switch to a fleet powered entirely by natural gas. The refueling is conducted at an outdoor location and will be constructed in compliance with the requirements of NFPA Standard 52. What fire safety requirements should be imposed in the service, maintenance, and storage areas? Some brief thought will show that the fire protection considerations for this situation are unlike any other common fire protection scenario and will not be found in any current standards or codes. Unlike gas or diesel fuels, the vehicle fuel is gaseous, lighter than air and under high pressure. Unlike fixed indoor storage tanks, the fuel cylinders move into, out of, and around the garage on a frequent basis.

Walls are of concrete masonry unit construction. The steel roof deck is covered by a built-up composite roof and is supported by open steel trusses which allow the free passage of air. Air supply is through ducted fan units. Exhaust is accomplished by roof mounted "upblast" fans.

DESCRIPTION OF NATURAL GAS

Natural gas is a term which is used for a hydrocarbon vapor having a number of constituent components, including methane, ethane, propane, butane, and other heavier hydrocarbons. The primary component is methane (CH_4), to the extent that the other components contribute only a very minor portion of the total mixture. The higher the percentage of methane, the more stable the heating value of the fuel becomes, and the higher the octane rating. Both qualities are desirable for fuels, therefore natural gas processes are typically designed to minimize or eliminate the non-methane components of the product. Because of the dominance of methane in the mixture, hereafter in this article the term natural gas shall be understood to mean methane.

Natural gas at standard temperature and pressure (STP) is lighter than air. When released in still air, it forms a buoyant plume and, if unobstructed, rises rapidly toward the ceiling.

Natural gas is flammable at concentrations of 5 percent to 15 percent in air. The lower flammable limit (LFL) of 5 percent is the fuel-lean limit and the upper flammable limit (UFL) represents the fuel-rich limit. Burning of the gas will occur only if an ignition source is in contact with a volume of gas which is between the LFL and UFL.

The hazard posed by a natural gas leak depends strongly on the mode of release. For example, the behavior of a pressurized natural gas release due to failure of a pressure relief device differs dramatically from the release of gas from a ruptured fuel line within an enclosure such as a fuel tank cowling. Petroleum industry studies suggest that, due to jet mixing effects, an unobstructed, high velocity, vertical gas release will disperse below flammable limits within a few feet of the discharging orifice when a sufficient supply of fresh air is available for entrainment.

Natural gas releases of sufficient volume to form a flammable cloud near ceiling level are best removed by ceiling level exhaust fans which exhaust to the outside of the building, above the roof. Optimal air flows in such situations rapidly remove the flammable mixture from within the building. The anticipated presence of a flammable cloud for some period at ceiling level during a significant release implies that ignition sources near ceiling level must be controlled or removed.

DESCRIPTION OF CNG FUEL

Compressed natural gas comes from one of the two following sources:

1. If taken directly from the municipal pipeline, its consistency varies from 85 percent to 99 percent methane, depending on supplier, with the balance of the gas being made up primarily of heavier hydrocarbons such as butane and ethane. The gas is odorized.
2. If taken from liquified natural gas, it is over 99 percent pure methane. The gas is unodorized unless odorant is added on-site.

In either case the gas is compressed to 3,000 to 3,600 pounds per square inch (psi) at room temperature for vehicular storage. For some fueling stations, natural gas storage pressures can reach 5,000 psi or higher. The compression process and refueling are assumed to occur outside the maintenance building (a typical arrangement) and are thus outside the scope of this article. Widely accepted fire safety requirements for these processes and their associated facilities are given in National Fire Protection Standard 52, "Compressed Natural Gas Vehicular Fuel Systems."

DESCRIPTION OF CNG FUEL SYSTEMS

CNG fueled buses closely resemble their conventionally fueled counterparts. Construction of the fuel system is constrained by the requirements of NFPA Standard 52. The gas is stored in manifolded series of cylinders typically made out of lightweight aluminum shell with fiberglass filament outer wrap. Fuel cylinders

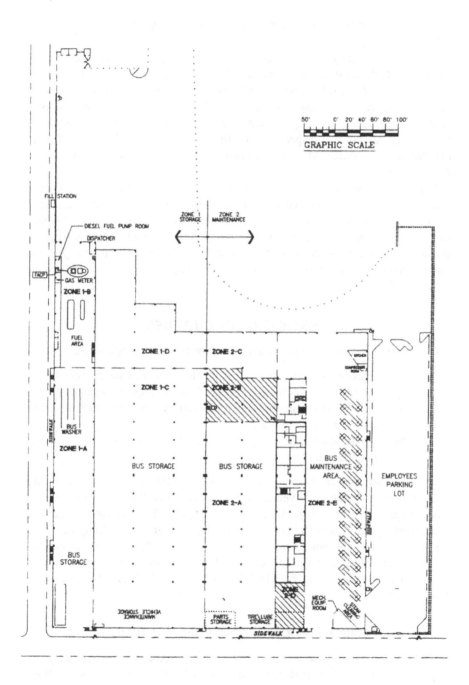

Figure 1. Typical maintenance facility.

may be located either on top of or underneath the body of the bus. A total storage capacity of approximately 15,000 standard cubic feet is typical for a 40 foot bus. Gas is supplied through stainless steel tubing to a pressure regulator assembly that reduces the 3,000 psig initial cylinder pressure to the approximately 18 psig required by the engine. The 18 psig gas is fed into a carburetor for use by the engine, beyond which point the system processes resemble those of a conventional spark-ignition engine.

The high pressure system for a typical CNG-fueled bus is shown in Figure 2.[1] For this model, twelve cylinders are mounted in a vented cowling above the bus. Each cylinder is capable of holding 1,324 standard cubic feet (scf) of gas when fully fueled to 3,000 psig at room temperature. Each cylinder possesses a thermally activated, non-resettable pressure relief valve and a manual shutoff valve. Cylinders are manifolded in three groups of four. Each bank of cylinders is isolated from other banks by an electric solenoid valve which is open while the engine is running, but which de-energizes to a closed state once the engine has been shut off for thirty seconds.

Fuel is supplied to the engine through a two-stage regulator arrangement located in a vented rear service compartment. The first stage reduces pressure from 3,000 psig to about 300 psig. The second stage then reduces the 300 psig fuel pressure to the 18 psig necessary for proper engine operation.

The engine compartment is supplied with a dry chemical fire extinguishing system which can be activated manually by the driver or automatically by high heat detection in the engine compartment. The primary purpose of this system is to increase the safety of on-board passengers against a bus fire originating in the engine compartment.

CNG HAZARD ANALYSIS

Natural gas can present an anoxia (asphyxiation) hazard when it displaces oxygen in a confined area without adequate ventilation.

Human error leading to release of natural gas during maintenance operations is the single most important risk factor identified. Improperly performed maintenance can lead to an immediate release in the maintenance garage or can contribute to leakage after the bus has left the facility. High quality maintenance is dependent upon proper training, procedure documentation, a safety conscious attitude among both management and employees, proper management support toward required changes in procedures, and audits to ensure that the program is working as intended.

The most severe fire protection hazard in a CNG fueled bus maintenance garage would be created by the leakage, accumulation, and ignition of a natural gas

---

[1] The Orion Bus is produced by Bus Industries of America, a subsidiary of Ontario Bus Industries.

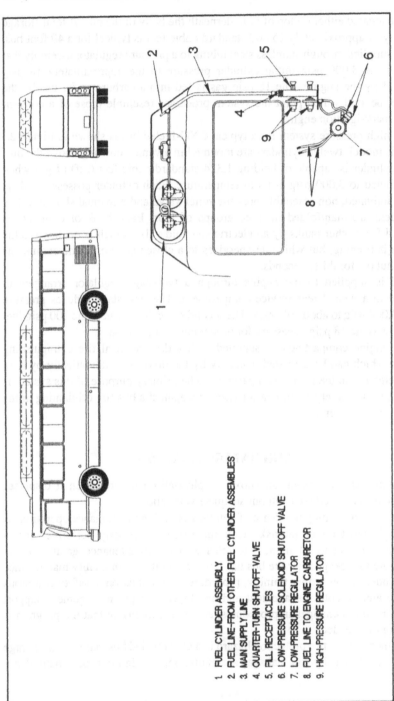

Figure 2. Fuel system schematic.

1. FUEL CYLINDER ASSEMBLY
2. FUEL LINE—FROM OTHER FUEL CYLINDER ASSEMBLIES
3. MAIN SUPPLY LINE
4. QUARTER-TURN SHUTOFF VALVE
5. FILL RECEPTACLES
6. LOW-PRESSURE SOLENOID SHUTOFF VALVE
7. LOW-PRESSURE REGULATOR
8. FUEL LINE TO ENGINE CARBURETOR
9. HIGH-PRESSURE REGULATOR

cloud from one or more pressurized fuel cylinders. Each of these factors can be examined separately to determine the manner in which they are likely to occur, drawing on recent applied research literature to help typify them [1, 2].

The elements of a fire hazard are realized when fuel and oxygen are present in sufficient ratios, and an ignition source is present. Combustible mixtures can be ignited by surfaces heated in excess of 1,000 degrees F, or by any but the smallest sparks (any above approximately 1/2 mJ; .48 mJ is the spark ignition energy for flammable concentrations of methane at atmospheric pressure, with electrodes 1/10-inch apart).

Based on existing studies of CNG vehicle fuel systems [1], natural gas releases can be characterized as one of four modes. First, gas can be released from small leaks on the order of 10 standard cubic feet per minute (scfm). Studies show that such leaks can occur in the high pressure system as the result of poor maintenance practices, such as assembling fuel system fittings which have been contaminated by dirt or grit. Second, intermediate level releases can occur with flows between 100 and 3,000 scfm due to the spontaneous failure of thermal pressure relief devices (PRD's) or serious maintenance error. In the Orion bus, a PRD release results in an unobstructed vertical gas jet with a maximum flow in this range. Third, a hypothetical worst case (guillotine) line break with a flow greater than 3,000 scfm has been examined using computer modeling [1]. The scenario allowing the occurrence of such a break inside a maintenance garage is difficult to construct, barring gross maintenance failure. Finally, gas can be released through the catastrophic rupture of a tank. Such a rupture could occur if the filament tank wrapping is sufficiently damaged by chemical or physical means. No ruptures of cylinders used for CNG buses is known. Limited data of rupture in similar CNG car fuel cylinders indicates that the rupture releases all gas almost immediately but does not necessarily result in ignition.

Computer modeling and thermodynamic analysis indicate that flammable concentrations of natural gas, when released from a pressurized CNG system operating in the range of 0 to 3,600 psig would be colder, but still lighter than, ambient air.

The behavior of gas leakage varies strongly with the size of the leak. Studies indicate that slow leaks (on the order of 10 scfm) dissipate below flammable limits within a few inches to a few feet of release [1, 2] and thus do not form a significant hazard of accumulation within a high-bay bus maintenance facility. Leaks significantly greater than 100 scfm are more difficult to quantify. Evidence exists to suggest that unobstructed vertical jets evacuating into large open spaces, such as might occur during a pressure relief device failure, will dissipate below flammable limits within a few feet of the release point [3]. If subjected to a local ignition source, such a leak could result in a torch fire, producing localized damage to the bus and surrounding equipment.

Computer modeling shows obstructed releases (such as would be experienced underneath a bus or within the cowling) rise rapidly toward the ceiling and have the capability to accumulate at ceiling level [1],[2] The accumulation of a flammable cloud of natural gas poses the most severe consequences, as it can result in an explosion, with accompanying danger to personnel and damage to property.

CNG RECOMMENDATIONS

Based on a detailed review of the general considerations mentioned above, recommendations can be made for our hypothetical facility to minimize the possibility and effects of both slow and fast leaks.

The hazard of slow leaks is already low due to the tendency of the gas to quickly dissipate below flammable concentrations. Normal levels of ventilation present in a maintenance garage are sufficient to prevent accumulation of slow leaks. To further reduce the hazard of such leaks, the following recommendations are suggested:

1. Ensure that vehicular fuel systems are constructed to comply with NFPA Standard 52, *Compressed Natural Gas Vehicular Fuel Systems.*
2. Ensure that all mechanics who work on the high pressure system receive the necessary specialized training. Special care should be taken during maintenance work involving disassembly and reassembly of the high pressure fuel system.
3. Provide procedural controls (particularly hot work permits) in maintenance areas for work around potential fuel leakage points.
4. As part of a fuel maintenance program, provide a program for leak testing fittings following maintenance to bus fuel systems. Bubble testing or chemical detection of leakage is suggested.[3]

The large scale release of gas, in quantities sufficient to pose a threat of accumulation, represents a lower probability but far higher hazard potential. Based on analysis of computer modeling and empirical data, such leaks would be expected to create flammable regions in a localized plume above the leaking bus and rise, due to their buoyancy, to form a cloud at ceiling level.

Recommendations to prevent or mitigate the effects of such an occurrence are to implement recommendations 1 through 3 above together with the following; jointly these will mitigate the formation and ignition of a natural gas cloud by providing effective emergency ventilation and by minimizing potential ignition sources:

---

[2] Thermodynamic properties cause high pressure natural gas to cool when expanded abruptly to room pressure. Calculations conducted by one of the authors indicates that the gas remains buoyant in air following this cooling process.

[3] Empirical evidence suggests that leaks will generally manifest themselves through water vapor condensation and noise before they reach a level at which they will pose a significant accumulation threat.

1. Since the greatest natural gas accumulation potential is near the ceiling, provide for ceiling level exhaust fans in areas where buses are present. Such fans should be capable of achieving at least six ceiling level air changes per hour under emergency conditions and should be of construction suitable for a Class 1, Division 2, Group D location (see Recommendation 4). Emergency ventilation should be capable of being activated by a methane gas detection system or by manual pull stations located near building exits. In general, the philosophy is to evacuate the methane layer at ceiling level as rapidly as possible, rather than attempting to dilute it below explosive levels inside the building. For this reason, if there are existing downward-blowing, ceiling-mounted air inlets, they are recommended to be shut off under emergency ventilation conditions. Providing for automatic garage door operation (i.e., opening) in the maintenance garage upon activation of gas detection or manual fire alarm systems will contribute to the effectiveness of emergency ventilation measures.

2. Since the most probable natural gas accumulation zone is directly above the buses and at ceiling level, replace all ceiling and truss mounted open flame heaters with alternative heaters that have rooftop heating elements or that use heating elements which operate substantially below 1,000 degrees F. Several types of heaters are suitable for natural gas fueled vehicle facilities.

3. Provide a reliable methane gas detection system at ceiling level in bus service, storage, and maintenance areas throughout the facility and provide manual pull stations near exits from areas where buses are present. Activation of the methane detection system or of a manual pull station should result in activation of emergency ventilation systems as described in Recommendation 1 and shutdown of non-emergency electrical equipment as noted in Recommendation 4.

4. Following a large scale release of natural gas from a bus, computer modeling indicates that the gas will tend to rise sharply and spread along the ceiling. Since the methane detectors take a finite time to react to the presence of the gas following exposure (typically 10 to 20 seconds), the possibility of a flammable cloud exists for a brief time interval prior to actuation of emergency ventilation and shutdown of nonclassified electrical equipment. It is not necessary to electrically classify the entire building [4]. However, to minimize the possibility of ignition during this period, classify the area adjacent to the ceiling as a Class 1, Division 2, Group D area as defined in Article 500 of NFPA 70, the *National Electrical Code.* Replace unclassified electrical systems within this area or relocate them to below the classified area. The depth of area below the roof deck assigned as classified space is dependent on engineering judgment; it was assigned in the cases analyzed as eighteen inches.

5. Provide a mechanism for shutting off all unclassified electrical apparatus and circuits present in the bus areas. This shutoff would typically be through

shunt trips of central, non-emergency distribution panels. Emergency power circuits must be individually analyzed as to their location and the possibility of their forming ignition sources. For example, emergency circuits located near ground level and away from immediate proximity to buses with top-mounted fuel tanks may be allowed to remain energized, whereas emergency circuits located near ceiling level might need to have their emergency status reassessed or be converted to classified status.

6. A probabilistic risk study [5] has identified human error as the chief potential cause of a significant CNG release. Provide procedural controls to substantially depressurize bus fuel systems (to about 300 psig) prior to bringing the buses inside the facility for fuel system maintenance.

DESCRIPTION OF LNG FUEL

The following are key points needed to understand liquified natural gas (LNG) behavior and releases:

1. LNG is a flammable cryogen, with methane being LNG's primary component (up to 99.5% methane). It is stored as a liquid at approximately –260°F for vehicular fuel systems. In storage, the natural gas exists in both the liquid and vapor phase. When the liquid expands to vapor at STP, the volume increases approximately 600 times. One gallon of LNG will convert to 467.5 ft.^{3} of vapor. Typical storage pressure for a vehicular fuel system is 60 psi.

2. As LNG is released and vaporizes, the density of the –260°F vapor is greater than that of the ambient air. As thermal energy is transferred to the gas, the temperature increases, the volume further expands, and the gas becomes more buoyant.

3. At approximately –180°F, the density of the gas is the same as that of the ambient air. Once the temperature increases to –30°F, the vapor density of the gas has dropped to 0.6 times that of air, near that of room temperature methane.

4. At temperatures above –30°F, the behavior of LNG vapor is essentially the same as that of CNG.

5. Although methane is colorless and odorless, water condenses from the surrounding atmosphere during vaporization, temporarily forming a whitish vapor cloud.

6. When a small release occurs, the LNG flashes directly to vapor. If the release is of sufficient quantity, it remains liquid, falls to the ground, and pools or spreads according to the physical conditions encountered.

7. LNG in contact with a surface will immediately begin to draw the available "sensible" heat from the surface and surrounding air. It will vaporize to

the extent that sensible heat is available; simultaneously, the surrounding surface is cooled. Soon, the energy transfer reaches a steady state, and vaporization is dominated by that which is caused by convection between the ambient air and the LNG pool.

8. The cloud formed by a vaporizing pool of LNG is initially cold and dense, and remains close to the ground. As this cloud warms due to the temperature of the surrounding air, it becomes lighter than air and rises.

DESCRIPTION OF LNG FUEL SYSTEMS

LNG fueled buses, like CNG fueled buses, closely resemble their conventionally fueled counterparts. Construction of the fuel system should be constrained by the requirements of NFPA Draft Standard 57, Liquified Natural Gas Vehicular Fuel Systems. The gas is stored in a cylinder or manifolded series of cylinders, called dewars, specially designed for the storage of cryogens. Each dewar by definition is a vessel having double-wall construction, with the intermediate space evacuated to enhance the insulating properties of the container. In some cases, the intermediate space may be filled with insulation material. Fuel cylinders are located beneath the body of the bus, primarily due to the weight of the storage arrangement and fuel. One benefit of LNG fuel storage systems is significantly increased fuel capacities over CNG systems of equivalent size. A total storage capacity of approximately 37,500 standard cubic feet is provided by a dewar containing 80 gallons of LNG.[4] As the liquified storage vaporizes and expands for engine use, gas is supplied through stainless steel tubing to a single pressure regulator that reduces the 60 psig dewar pressure to the approximately 18 psig required by the engine. The 18 psig gas is fed into a carburetor for use by the engine, beyond which point the system processes resemble those of a conventional spark-ignition engine.

Each cylinder possesses a thermally activated, non-resettable pressure relief valve and a manual shutoff valve. Cylinders may be manifolded to increase the total fuel stored when physical limitations prevent the installation of larger dewars. Each manifolded dewar is isolated from other dewars by manually operated shutoff valves. Dewars are also equipped with excess flow check valves, which limit the maximum flow rate of fuel from a dewar to a level sufficient to meet engine demands, plus some safety factor. The excess flow check valve is intended to mitigate the hazard presented by a major fuel line failure such as the line being severed or significantly ruptured.

As with CNG fueled vehicles, the engine compartment is supplied with a dry chemical fire extinguishing system.

[4] This compares to the 15,000 scf maximum capacity for CNG fueled buses 40 feet long.

LNG HAZARD ANALYSIS

In addition to the hazards presented by CNG, which are duplicated by an LNG release after the vapor's temperature warms to approximately –30°F, an LNG release presents distinct hazards during the time the fuel is vaporizing and warming from –260°F to –30°F.

1. LNG is a cryogen. Cryogenic hazards are presented by the LNG, by cold vapors, and by any surface in contact with the initial stages of an LNG release. Personnel contact with any of these cold sources may result in cryogenic burns resulting in permanent tissue damage. The physical properties of equipment and structures may also change (e.g., become brittle) as a result of contact with cold sources.
2. The cloud produced upon initial vaporization is heavier than air. As a result, flammable mixtures of gas vapor and air are likely to accumulate near the floor until such time as the vapor warms to the point of buoyancy and begins to rise. Ignition sources near the floor as well as the ceiling must therefore be controlled.
3. The dense vapor cloud initially developed is subject to drift due to normal air movement patterns within the building. As a result, the location of the flammable cloud at the ceiling may not be directly over the release, as it will be affected by air movement patterns, and physical conditions within the building.

LNG RECOMMENDATIONS

As discussed previously for CNG, recommendations can be made for slow and fast leaks of LNG. Since the slow leak of LNG is expected to occur in the gaseous state or immediately flash to vapor, recommendations previously made for slow CNG leaks remain valid. To mitigate the hazard presented by slow LNG leaks in service pits, the concepts of gas (methane) detection, electrical area classification and shut-down of non-essential electrical circuitry should be extended to specifically include the service pit areas.

Fast leaks of LNG, conversely, required consideration of more than the extension of the CNG recommendations. For instance, as LNG releases may result in liquid as well as cold, dense vapor discharge, consideration must be given to locations where such materials may accumulate, even if the accumulation is expected to be short-lived. Issues of containment, drainage, and temperature of LNG in the liquid state must be addressed. Issues of vapor removal, ignition source removal, and work procedures must be considered for cold LNG vapor clouds.

Electrical Area Classification should be applied to areas near, at, or below ground level where vehicle fuel system maintenance will occur and where flammable mixtures may collect. Electrical Area Classification should be Class I,

Group D, Division 2 in the affected spaces. The use of portable power tools should be closely controlled by work procedures established based on the hazards presented by LNG.

Ventilation system requirements for LNG maintenance facilities must consider that at certain periods of time after a release, the vapor (and therefore the flammable mixture which includes that vapor) is heavier than air and will fall to the lowest available elevation. Ventilation (exhaust) must be provided at low areas where cold natural gas vapor may collect.

For heating systems, the philosophy of avoiding open flame heating units in areas where flammable mixtures might collect needs to be extended to LNG-specific portions of the building, such as the mechanical service pits below grade.

The practice of minimizing the LNG release by the closure of all but one cylinder/vessel manual shut off valve should be applied to LNG vehicles when fuel systems work will occur. At such times, hot work procedures considering the hazards of LNG should be established. Normal work procedures should be reviewed to accommodate the differences in behavior of LNG vapors from diesel and gasoline vapors, while also understanding their similarities.

FIRE PROTECTION PHILOSOPHIES AND MEASURES

As with any hazard, the fire protection mechanism used to control the specific hazard presented by CNG and LNG fueled vehicles considers some very basic principles:

1. Identify the hazard (fuel).
2. Determine factors contributing to hazard probability and severity, and review methods to reduce the impact of those factors.
3. Develop a strategy for personnel safety and property protection including detection, suppression, and notification, each as appropriate.

The hazard presented is clearly the uncontrolled release of methane gas inside the maintenance garage.

Factors contributing to hazard severity include total amount of fuel which might be released, vehicle and building features contributing to gas accumulation, and potential sources of ignition. The maximum expected fuel release can be manipulated to some extent via bus design considerations of size and number of fuel storage tanks, and location, type, and number of fuel system safety devices such as shut off valves, check valves, pressure relief devices, and excess flow check valves. Maximum release quantities can also be controlled through defined maintenance procedures which limit the acceptable fill level a bus may have before entering the garage. Building features which can contribute to gas accumulation include building size, compartmentation at ceiling level and location and type of HVAC devices. Vehicle features which can contribute to gas accumulation include location of fuel storage, location of engine exhaust and PRD discharges, location of fuel system manual relief valve discharges, enclosures about the fuel system components (i.e., cowlings), and locations of fuel system regulators. Ignition sources within the maintenance garage can be controlled through electrical area classification, emergency shut down of non-essential electrical systems and components, established personnel safety policies such as "No Smoking" within the garage, and appropriate hot and normal work procedures specifically developed considering CNG and LNG hazards.

Personnel safety and property protection methods include a number of hazard options such as: automatic smoke detection, heat detection, flame detection, methane (gas) detection located at low levels in the pits or at ceiling level, liquid level spill detection for the pit areas, and specific automatic identification of CNG/LNG vehicles when they enter the garage. Additionally, options exist for on-board smoke, heat, or gas detection for each vehicle. Options for suppression systems include automatic sprinklers, standpipes, foam-based systems, alternative agent (i.e., halon replacement) systems, dry chemical systems, and portable extinguishers. Personnel notification may be accomplished via audible signal, visible signal, manual paging, or recorded message systems tied to the automatic detection provided. Remote notification of Emergency Response personnel can be tied to the automatic detection. Interface may also be provided between automatic detection and suppression and emergency ventilation systems, as well as Emergency Shut Down (ESD) procedures for the vehicle and for the garage.

IMPORTANCE OF PREVIOUS INCIDENTS
AND OPERATING RECORD

As in all emerging technologies, the best available knowledge is collected to develop concepts, apply conceptual theories to design applications, and test the resultant product, process, or guideline as a means of providing validation of the original (or modified) concept. To develop a practice for establishing the fire and life safety of a maintenance facility for natural gas fueled vehicles, it is particularly valuable to analyze information related to specific incidents of fuel release, product failure, and fire related to natural gas fuels. Proper analysis of operating experience will provide a more useful record of performance than is currently available. Based on this analysis, design efforts can be more precisely tailored to address fire and life safety issues for these facilities.

SUMMARY

Through this two-part series, the authors have illustrated a general approach for a practicing design professional faced with a complex, novel fire protection problem. In the first chapter, a process has been described indicating how to develop a set of practical fire safety requirements for a situation which lacks any direct, preexisting code guidance. The process includes creating a hypothesis or hazard analysis, conducting a broad literature and information search, assessing the gathered information for applicability to the given problem, and developing appropriate recommendations based on the new-found knowledge. Over time, recommendations will be refined through analysis of operating experience. However this process can be accelerated through additional research, including the properly validated use of computer modeling. In this second chapter, an example of the basic process has been illustrated for modification of a maintenance facility to accommodate natural gas fueled vehicles.

REFERENCES

1. SAE Technical Paper 922486, *Extent of Indoor Flammable Plumes Resulting from CNG Bus Fuel System Releases,* Michael J. Murphy, Susan T. Brown, and David Philips, Battelle, 1992.
2. SAE Technical Paper 931814, *Gaseous Fuel Transport Line Leakage—Natural Gas Compared to Hydrogen,* M. N. Swain, Analytical Technologies, Inc. and G. J. Schade and M. R. Swain, University of Miami, 1993.
3. API Recommended Practice 521, *Guide for Pressure-Relieving and Depressurizing Systems* (2nd Edition), September 1982.
4. *Case Study: Electrical Classification for a Natural Gas Vehicle Maintenance Facility,* Ralph Kerwin, Jack Woycheese, Gage-Babcock & Associates, Inc., April 1993.
5. Gas Research Institute, *Risk Assessment of Indoor Refueling and Servicing of CNG-Fueled Mass Transit Buses, Phase II, Final Report, Volume I,* November 1990.
6. *Fire Protection Requirements Based on the Use of LNG and CNG Fueled Vehicles at the Seminary Facility Maintenance Garage,* Thomas J. Forsythe, Jack Woycheese, Gage-Babcock & Associates, Inc., September 1994.

CHAPTER 4

Characteristics of Natural Gas Leaks in Bus Garages

Pasquale M. Sforza and Herbert Fox

With the growing use of compressed natural gas (CNG) in fleet vehicles, it is of considerable importance to understand the characteristics of this fuel as it may leak into enclosed spaces forming a flammable plume. The purpose of the current work is to provide insights, based on computational simulations and experimental data, into the fluid mechanics and thermodynamics of CNG plumes. We describe the behavior of a leaking CNG system in terms of both jet-like momentum-dominated, and plume-like buoyancy-dominated, effects. This can help establish bounds on the needs for safety devices in an enclosed space. Also presented is a potential experimental setup that may be used to verify future theoretical analyses and become an important source of data for the CNG using community.

During the course of the next several years, due to pressure arising from changing and ever more stringent emissions standards, there will be increasing reliance on alternatively fueled vehicles. A significant focus of these alternatives will be natural gas, in both its liquid (LNG) and compressed (CNG) forms. In the short term we can expect that major fleet operators, like the U.S. General Services Administration, utilities, services industries, and buses, will increasingly depend on both LNG and CNG as their primary fuel [see 1-4]. There remain, however, fire safety issues related to such conversions. These center on the effects of potential natural gas leaks from high pressure fuel tanks and subsequent rates of dispersion of the resulting plume in enclosed storage and maintenance areas. The question of instrumentation placement and response time to detect such leaks needs to be evaluated. The effectiveness of current and proposed heating, ventilating, and air conditioning systems (HVAC) coupled to alarming devices, needs to be assessed to determine their potential in clearing the facility of such leaks. The current study first reviews the literature of recent computer-based studies of plume dispersion in garages and other enclosed areas, presents practical

43

engineering plume and dispersion calculations for typical garage conditions, and offers a protocol for affordable laboratory testing of gas dispersion in scaled models of bus storage buildings.

BACKGROUND AND PREVIOUS STUDIES

Over the course of the last several years there have been several studies of CNG dispersion in enclosed spaces. In the earliest of these, Karim focused on automobile usage and, with only minimal operating experience with CNG-fueled vehicles, concluded that the use of methane would not pose any additional hazards to automobile use than does regular gasoline [5]. This effort and a later one by the same author, draw their conclusions based on a study of the physical properties of CNG compared to other fuels [6]. Some of these conclusions are as follows:

- CNG has relatively high flammability limits, yet its flammability range is sufficiently narrow compared to other fuels. It takes a minimum of about 5 percent, by volume, of methane in air at ambient conditions to just support continuous flame propagation. This compares to about 2 percent for propane and 1 percent for gasoline vapor.
- Combustible mixtures involving methane tend to be less tolerant than other fuels to the presence of diluents rendering fire fighting and suppression a relatively easier task.
- The minimum energy required to effect a successful ignition of a fuel-air mixture within the flammability range is relatively large for mixtures involving methane. Such mixtures therefore are much more difficult to ignite than other fuels.
- Methane is safer than gasoline and propane with regard to denotation.
- The low molecular weight and low specific gravity assist in dispersing the fuel quickly should a leak occur.
- The nature of the CNG storage tank design and quality of construction suggests that they will likely be safer than their existing counterparts for gasoline or diesel fuel.

Not that there are no hazards. There are, and these were also identified:

- The relatively low auto-ignition energy (although comparable to that of gasoline fuels) of methane can be a problem. Ignition is usually assured in the presence of thermal sources such as sparks, lighted matches, hot surfaces, and open flame heaters. As a consequence these must be considered carefully in facility design and modification.

- The low molecular weight of methane, while assisting in plume dispersion, may permit the gas to linger in confined spaces near the ceilings in many garages.

The Sacramento Regional Transit District, like other agencies, is converting their fleet of buses to CNG. Two studies of theirs are relevant and both cite the relative absence of historical safety data for CNG bus operations due to the recent adoption of this technology [7, 8]. Basic safety conclusions, however, are largely the same as those above; that is,

- Facility modifications for the use of CNG relate largely to increased ventilation requirements. These derive from flammability limits and the perceived need to exhaust CNG as quickly as possible.
- Since vehicle storage is outside in Sacramento, only few changes are required to the maintenance facility.
- Improved gas detection, ventilation, and emergency shutdown systems are required in the vehicle storage and maintenance areas. Adequate ventilation is defined as that which is sufficient to prevent accumulation of significant vapor-air mixtures in concentrations over 25 percent of the lower flammability limit.

Indeed that last point is the key to safe operations with CNG. The monitoring, alarm, and HVAC systems must be designed to quickly identify that a leak has occurred, that personnel and systems are notified, and that the ventilation systems are operated to clear the facility as rapidly as possible, at all times maintaining low concentrations of methane.

Three recent studies of CNG dispersion in enclosed areas are particularly relevant [9-11]. All use large scale computer fluid dynamics models to assess CNG dispersion and safety. It is important to emphasize that all results were developed from the computer models with no experimental backup or verification. The focus of the analyses are effects of the low flammability limit coupled to the low molecular weight/high diffusion properties.

With regard to vehicle tunnel studies and based on the computer modeling [9]:

- The analysis indicates that the released natural gas spreads along the ceiling of the tunnel and is not a confined cloud.
- Such an unconfined cloud cannot detonate with ordinary ignition sources, and therefore a CNG explosion in a ventilated tunnel is highly unlikely.
- Gas ignition, should it occur, releases thermal energy but would not produce an explosion blast wave.
- Longitudinal air movement, which normally exists in tunnels, of even small amounts reduces fire hazards significantly.

As a consequence, the overall hazard of CNG vehicles was assessed to be not greater, and may actually be lower, than typical gasoline/diesel vehicles in a tunnel environment. Of particular importance is the nature of the tunnel ceiling configuration. In most tunnels this is largely unobstructed by ventilating ducts and other utility services. In addition, exhaust air registers are located flush with the ceiling making CNG removal after a leak relatively straightforward because of the rapid diffusion.

The purpose of the study by Grant et al. was to evaluate the hazard associated with parking CNG fueled vehicles in large garages [10]. Their concern, as that of the current study, arises from the fact that the enclosed garage might inhibit gas dispersion. They make use of large-scale computer fluid dynamics programs without experimental data or backup to arrive at the following conclusions:

- Small CNG leaks present the same hazard whether the vehicle is in a garage or in the open. Although a small CNG leak (with order of magnitude of $10 \text{ft}^3/\text{min}$) could be ignited by a source close (within a few inches) to the leak, this hazard is less than the persistent and more widespread hazard of a gasoline leak.
- It is important to minimize ceiling pockets which arise because of the placement of hanging beams, ventilation, and other service ducts. While the gas may be briefly trapped in these pockets, the computer model suggests that diffusion continues to remove it.
- The quick rise of the gas (rapid diffusion) to the ceiling removes it from many of the potential ignition sources, although ignition from (unspecified) defective lights or wiring cannot be precluded.
- In leaks which arise from pressure relief device (PRD) failure, the computer model predicted no gas build-up in the overall garage, although ceiling pockets did trap the gas and large concentrations did occur. It is critical to have the placement of the exhaust registers at ceiling height to assure gas removal.

As a consequence, it was concluded that CNG vehicles pose no extraordinary risks inside a well-designed and well-ventilated public parking garage. It may be emphasized here that no rigorous definition of "well-ventilated" is provided by Grant et al., or for that matter in any of the literature cited here [10]. A practical definition may turn out to be one of the major tasks in the development of safe storage facilities for CNG vehicles.

The final study considered assesses the extent of flammable plumes resulting from CNG bus system leaks [11]. It should be noted that Murphy et al. consider CNG buses with tanks under the bus chassis, as opposed to buses which have the CNG storage on top of the vehicle [11]. The exhaust registers for the ventilation system in the garage modeled are located at ceiling height. It is pointed out again that the study uses a computer fluid dynamics program to develop conclusions without experimental data verification.

For their fast leak (venting of the entire tank in ten minutes) scenario:

- the flammable mixture spreads in a layer directly at ceiling height;
- some gas can approach the building walls but usually at less than flammability concentrations;
- since the flammable layer develops so quickly, human reactive control of leak consequences may be difficult;
- if the control strategy requires the activation of additional exhaust fans upon receipt of a gas detection signal, response times of the detector and fan system need to be considered;
- the model concludes that the existing HVAC system can clear the plume within about ten minutes or so.

For their slow leak (0.5 g/s to 2.0 g/s) scenario:

- A flammable plume exists alongside the leaking bus;
- plumes therefrom are subject to ignition from nearby sources.

Furthermore, it is concluded that additional modeling is necessary to determine effects of differing building configurations, ventilation rates, and exhaust register placement.

The most recent work on facility assessment addressing the same issues as those noted above is provided by Forsythe and Kewin [12]. The conclusions are quite similar to those previously cited, but again, much of this work is based on theoretical analyses and not experimental data.

Based on these efforts, it may be concluded that fast CNG diffusion rates work in favor of the safety of CNG-fueled vehicles. However, it is critical to understand the nature of the ventilation systems and their operations. This is especially true for those configurations of garages or enclosed spaces which can provide pockets for the gas, and which have dropped ventilating ducts and other piping and tubing which can obstruct the HVAC system.

FLUID MECHANICS OF NATURAL GAS LEAKS

A jet of gas escaping from a pressure line or vessel is propelled by the pressure difference between its container and the surroundings. The force due to this pressure difference is manifested as an increase in the momentum of the gas as it exits the confinement of the vessel. Once free of solid boundaries the momentum of the jet is affected only by the external forces that act on it. In free surroundings the only external forces that can influence the jet are gravitational, or buoyancy, forces. For natural gas whose density is only about 60 percent that of air, the surroundings will force the jet to rise. Thus, a natural gas leak has characteristics both of pure jets, which move through the surroundings because of an impulse provided by the original pressure differences, as well as those of pure plumes, which move through the surroundings because of buoyancy forces, like smoke rising from a cigarette.

Buoyant Jet

A general review of buoyancy-induced flows is presented by Gebhart et al. [13]. The specific characteristics of buoyant jets and their behavior in the pure jet and plume extremes have been carefully studied experimentally by Papanicolaou and List [14]. They focus on vertically directed buoyant jets and verify much of the previously developed theoretical and experimental relations for such flows [see 15]. The basis for analyzing buoyant jets rests on the character of the extreme cases:

- In a jet, momentum flux is preserved with no effect of density variations upon the motion; that is, the effects of the density differences are small compared to the momentum-induced forces and, thus, may be neglected.
- In a plume, the density deficit is preserved and the buoyancy force continually affects the motion; here the momentum-induced forces are the same order of magnitude as those developed by the density deficits.

In order to specify the magnitude of the various constants arising from the theoretical framework, Papanicolaou and List carried out extensive experiments on the mean and turbulent properties of a range of buoyant jet cases. The effect of buoyancy on jet flows is characterized by the initial value of the flux Richardson number, defined as the ratio of buoyancy forces, caused by the density variations, to inertia forces. For the conditions under study here it may be written as:

$$R = Q \, B^{1/2} / M^{5/4} \qquad (1)$$

where Q is the initial volume flux,

$$B = g[(\Delta\rho)_o/\rho_a] \, Q \qquad (2)$$

is the specific buoyancy flux and

$$M = QW \qquad (3)$$

is the specific momentum flux. The other variables are

ρ_a = density of the ambient atmosphere
ρ_o = density of the gas jet
$(\Delta\rho)_o$ = $\rho_a - \rho_o$ the density deficit
g = acceleration of gravity
W = initial jet speed in the vertical (z) direction.

Based on the characterizations noted earlier, the Richardson number is very small for jets, where the density variations are small compared to the momentum or inertia forces, and around unity for plumes where these forces are the same order of magnitude.

A characteristic length scale may also be defined as

$$\ell_m = M^{3/4}/B^{1/2} \tag{4}$$

to serve as a measure of the relative importance of the momentum and buoyancy fluxes, as shown by Fischer et al. With z taken as the distance in the flow direction Papanicolaou and List clearly demonstrate that for distances $z/\ell_m < 1$ the flow behaves like a jet while for $z/\ell_m > 5$ the flow behaves like a plume. In the intervening distances the flow makes a smooth transition from one type of flow to the other.

In general then, an escaping buoyant gas acts like a jet near the point of escape and tends to act like a plume as distance from the leak grows large. Detailed behavior of the velocity and gas concentration field is summarized by the relations provided in Appendix A. For the present purposes it is sufficient to visualize the flow, whether jet-like or plume-like, to be a slender cone-like region of included angle of about 15° within which the gas is continually ascending and diluting by the entrainment of the ambient air. If there is a barrier, like a ceiling, the flow will spread out over it, radially diverging from the point of impact. It is important to emphasize that the ultimate trend of the flow is to become plume-like since, at large distances from the leak rate, z will eventually become larger than $5\ell_m$. This is true even if the original direction of the escaping jet is far from vertically upward.

Characteristic Lengths and Leak Rates

The most convenient measure for characterizing the behavior of the gas leak is the characteristic length ℓ_m given by Eq. (4), which becomes

$$\ell_m = W^{3/4}\, Q^{1/4}/3.59 \tag{5}$$

when ℓ_m, W, and Q are expressed in dimensions of feet, ft/s, and cubic ft/s, respectively. Now assume that the gas may by taken to be methane with a density 60 percent that of air. The variation of ℓ_m with Q (in SCFM) for various values of the initial jet velocity W is shown in Figure 1. For reference purposes, also shown in the figure are a number of leak rate scenarios for various maximum initial rates[1] and may be classified as damaged ferrule, contaminated fitting [11], PRD release [17], small tube break, and large tube break. The means by which these were determined are shown in Appendix B.

---

[1] Note that those initial rates decay exponentially with time as the gas cylinder empties.

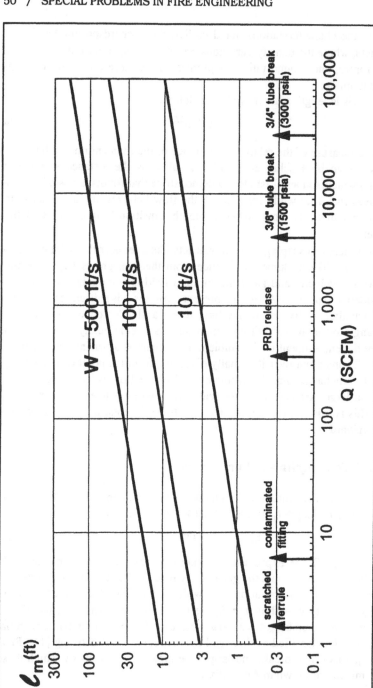

Figure 1. Characteristic length scale, ℓ_m, for various initial velocities, W. Typical leak rate scenarios indicated for reference.

Many full-size (40 foot) buses have their cab tanks installed in a roof-mounted configuration. The geometry of a buoyant jet release from such a tank is shown schematically in Figure 2. The distance from the roof to the ceiling is taken as nominally 10 feet, typically of bus storage garages. Then for all cases where ℓ_m is 10 feet or greater the (vertical) gas release will be jet-like, while for all cases where ℓ_m is 2 feet or less the gas release will be plume-like. For cases between these values the buoyant jet will make the transition from jet-like to plume-like flow. The graph in Figure 1 has been redrawn in Figure 3 to illustrate these regions.

Also shown on Figure 3 are contours of constant concentration of methane, drawn for values of 5 percent, representing the lower flammability limit, and 2 percent, representing 40 percent of the lower flammability limit. The contours are for the mean centerline (maximum) concentration at a distance 10 feet above the leak point, that is, nominally at ceiling level. Because the buoyant jet flow is turbulent, instantaneous values of gas concentration fluctuate above and below the mean value, so it is prudent to consider a factor of safety, here equal to a value of 2.5. The selected value of 2 percent is typical of alarm levels used in other CNG applications [3], although other researchers [18] suggest that a factor of 2 may be sufficient.

These contours indicate that small leaks, like those due to contaminated or damaged fittings, should become diluted to a safe level by the time the gas reaches the ceiling. Major leaks, like those due to a ruptured gas line, are likely to spread out across the ceiling prior to achieving safe dilution levels. On the other hand, moderate leaks, like those from the activation of a PRD, achieve safe dilution at some intermediate location. It should be noted that the tendency for the buoyant jet to rise makes typical domed skylights a trap for the gas, as illustrated in Figure 2. These structures may require individual venting if they are not naturally leaky enough to permit easy gas escape. For cases where the release is not vertical, for distances along the jet axis less than ℓ_m the flow will be jet-like. When the distance is around 5 ℓ_m buoyancy will begin to dominate and the jet will be moving more and more vertically, see Figure 4. A discussion of methods for dealing with such cases may be found in [13].

RECOMMENDATIONS FOR SCALED EXPERIMENTAL STUDIES

Experiments on the dispersal of a gas leak of given magnitude in an enclosed space may be carried out in situ. Although such tests provide information on the behavior of the gas under normal operating conditions, they do not offer a practical means to evaluate the effectiveness of a variety of corrective measures which might be employed to enhance dilution and diffusion of the escaped gas to safe levels. The purpose of this section is to outline a possible experimental scheme and show the types of results that may be obtained.

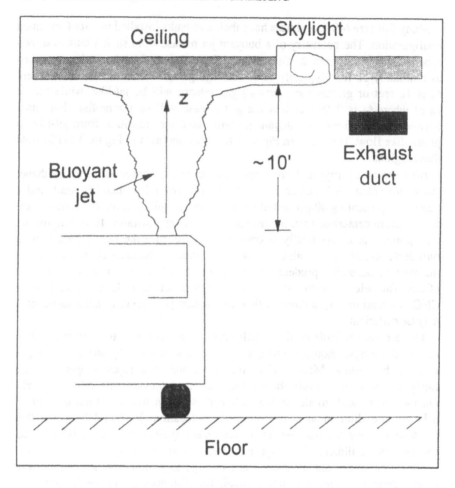

Figure 2. General schematic of gas leak: jet spreading over ceiling, gas trapping by skylight and typical relative position of exhaust duct.

In order to evaluate contemplated safety measures it is common to employ an appropriately simulated experimental facility of manageable (laboratory) scale. A physical model of this type serves as an analog of the full-scale facility and permits intensive experimentation on gas movements under a multitude of scenarios within a time and cost envelope of generally acceptable proportions. In addition, results of such experiments yield a database which may be used to validate computational fluid dynamic simulations that are useful in other design and analysis applications of a similar nature.

Any experimental program will have several steps, independent of the facility being modeled. These are discussed next.

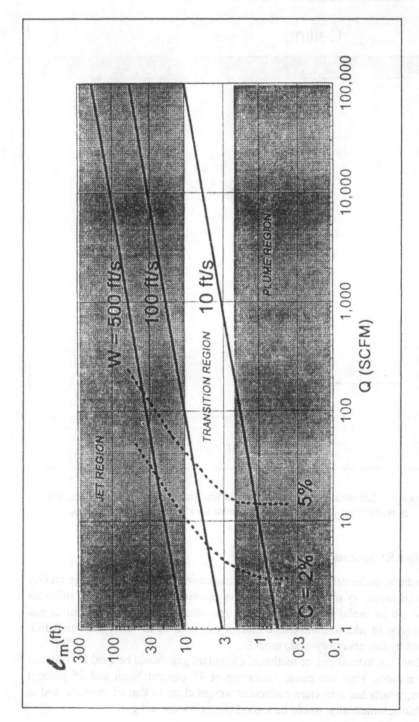

Figure 3. Contours of 2 percent and 5 percent mean gas concentration on the centerline at a distance of 10 feet from the release point. Regions of jet behavior for distances up to 10 feet are shown for reference.

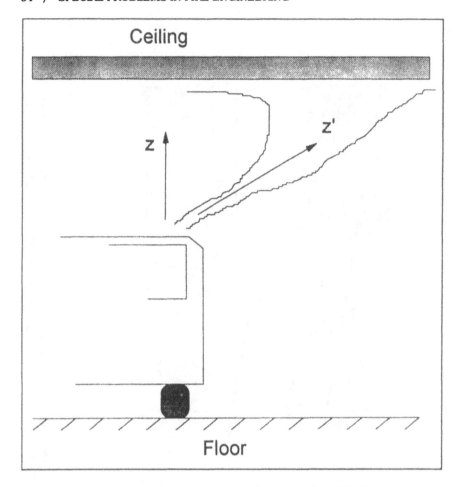

Figure 4. Schematic diagram of non-vertical buoyant jet showing initial jet behavior and ultimate plume behavior as z' grows larger than $5\ell_m$.

Baseline Experiment

In order to understand the general gas dispersion patterns in the storage facility it may be necessary to carry out a baseline experiment at the site. This information would be useful in itself because it provides documentation of actual gas movements. In addition, the data collected can be used to verify the simulation afforded by the laboratory-scale model.

Rather than natural gas, or methane, a simulant gas should be used for obvious safety reasons. First run neon, consisting of 75 percent Neon and 25 percent Helium, which has a mixture molecular weight close to that of methane and is available commercially, would be a good choice for the test gas.

Simulant gas releases in the full-scale tests would yield the following sorts of information:

- visualization of the gas leak flow patterns
- global leakage characteristics of the building
- performance of the supply and exhaust system
- flow characteristics at open doors and air curtains
- flow characteristics of the ceiling configuration, e.g., skylights, ducts, etc.

With the overall flow characteristics of the building and the ventilation capability of the existing systems with respect to gas leaks reasonably well-characterized, then, the model design may be initiated.

Model Design and Tests

Based on study of the existing garage configurations and other, similar model tests, a reasonable scale for the model seems to be 1:48. It is unlikely that a smaller model will be suitable; a larger model becomes unwieldy. Scaled buses can be readily obtained. With the scale as selected it is practical to model all significant features with a characteristic dimension, in full scale, of 6 inches or more; this yields a model dimension of 0.125 inch. The other major model factors are the porosity (leakage) on the walls, roof, and air curtain doors, as determined from the preliminary full-scale experiments.

A large number of the tests performed in the full-scale case would be repeated in the model. The results of these experiments would be compared to the baseline full-scale study in order to verify the validity of the simulation.

Once confidence in the model is established it will be possible to use it to evaluate methods to enhance dilution of potential gas leaks and then used, for example, to establish necessary program logic for controlling the building HVAC systems. Some possibilities may be noted at this time. It is likely that program logic will be a function of the location of the CNG leak site and a function of the type and location of CG sensor systems. Thus, the strategies that may evolve will necessarily be different depending on the way the leak may be totally characterized. Clearly, listing all the possibilities is premature at this stage, and is more appropriate once a good quantitative understanding of the actual flow patterns is in hand.

Full-Scale Experiments

Based on the model data, a number of locations which are representative of those studied would be identified and corresponding full-scale tests carried out. The purpose of these tests would be to quantify gas concentrations as a function of time at a point near the ceiling and directly above the release and validate the recommended program logic. Similar measurements at several radial distances from this central point will also be performed. The objective here is to quantify the dilution rate under normal conditions. As a corollary to these measurements the gas concentration as a function of time at some of the ceiling unique configurations would also be recorded.

CONCLUSIONS

The consensus of the literature reviewed is that natural gas poses no greater hazard than conventional fuels in enclosures, although its characteristics require some different safety considerations. Of special importance are the ventilation characteristics of the enclosure, the provisions for leak detection, and the operation of the ventilation system to effect safe and rapid gas evacuation. All of the information thus far reported is based on anecdotal operational experience and, in some cases, computational modeling of specific scenarios. There has been no systematic experimental evaluation of the problem to either develop empirical approaches or validate computational models.

As a step toward rectifying this lack we have presented here rigorous scaling criteria for high pressure natural gas leaks. These are based on a substantial background of experimental data in the general fluid mechanics literature. A method is presented whereby contours of gas concentration may be calculated in order to determine regions of flammability for various leak rates. These results can also be used as test cases for computational simulation codes.

An outline for developing properly scaled laboratory simulation of gas leaks in full-scale garages has also been presented. Such investigations entail some critical measurements in the full-scale facility followed by verification in the model. Once such agreement is obtained the laboratory model can be utilized to carry out the systematic experiments necessary to assess the validity of both empirical and computational models for facility safety.

APPENDIX A
Flow Field Relations

The results of a wide range of experiments led Papanicolaou and List [14] to the following best fits for the centerline properties for vertical buoyant jets:

$$\frac{M^{1/2}}{zW_c} = 0.132, \quad \frac{SQ}{zM^{1/2}} = 0.147 \; for \, jets, \, i.e. \, for \, \frac{z}{\ell_m} < 1 \qquad \text{(A-1)}$$

$$\frac{M^{1/2}}{zW_c} = 0.260 \left(\frac{z}{\ell_m}\right)^{-2/3}, \frac{SQ}{zM^{1/2}} = 0.070 \left(\frac{z}{\ell_m}\right)^{2/3} \; for \, plumes, \, i.e. \, for \, \frac{z}{\ell_m} > 5 \quad \text{(A-2)}$$

Here W_c is the mean velocity along the centerline while $S = C_o/C_c$ is the ratio of the initial gas concentration at the leak point to the mean concentration along the centerline. The width of the profiles of mean velocity and concentration are given respectively by

$$\frac{b_w}{\ell_m} = 0.108 \left(\frac{z}{\ell_m}\right)^{0.995}, \frac{b_c}{\ell_m} = 0.129 \left(\frac{z}{\ell_m}\right)^{0.994} \tag{A-3}$$

where b represents the radius at which the velocity (subscript w) or the concentration (subscript c) drops to 1/e of the centerline value. It is clear that the concentration field is somewhat wider than the velocity field since

$$\frac{b_c}{b_w} \approx 1.19 \tag{A-4}$$

Relations are also given for other properties such as mean flow profiles and turbulence properties such as velocity and concentration fluctuation intensity as well as turbulent transport properties, but these are not reproduced here [14].

APPENDIX B
Leak Rates

A range of leak rates may be determined from an examination of the physical arrangement of the tank storage and plumbing. First we assume that the tanks themselves are fail-proof. This is commonly accepted in the field and is mentioned frequently in the literature surveyed in Section 2. The leaks of concern here, therefore, arise from tubing breaks, PRD releases, fitting malfunctions, etc. The largest leaks will be due to tubing failures at the maximum operating pressure, typically 3000 psia. There are generally three major tubing sizes in bus fuel supply lines: 0.75-, 0.50-, and 0.375-inch diameter. Assuming the line is cleanly severed, the maximum leak rate may be determined by considering the leaking gas escaping as a sonic jet and expanding into a free air environment at standard atmospheric pressure (14.7 psia) and a storage facility air temperature of 50°F. The gas in the tanks may be assumed to be at the 50°F room temperature as well, and to be composed entirely of methane. The maximum flow rate is given by

$$Q_{std,o} = 19.16 p_{t,o} d^2 \tag{B-1}$$

where $Q_{std,o}$ is measured in SCFM, the initial tank pressure, $p_{t,o}$, is in psia, and the diameter of the leak hole is in inches. It is important to emphasize that this relation gives the initial leak rate only; the leak rate will decay exponentially with time as the gas escapes. The pressure in the tanks drops according to

$$p_t/p_{t,o} = exp\,(-0.219 d^2 t) \tag{B-2}$$

where t is in seconds. The mass of gas ejected up to any time t is

$$m(t) = m_o (1 - p_t(t)/p_{t,o})$$ (B-3)

where m_o is the mass of gas initially in the tanks. Considering the cases of full tanks ($p_{t,o}$ = 3000 psia), half-full tanks ($p_{t,o}$ = 1500 psia) and a variety of tube or hole sizes leads to the following table of maximum (initial) leak rates and approximate times (in minutes) necessary to eject 99 percent of the mass originally in the tanks (Table B-1).

Note that the tubing size of 0.75 inch exists only as the manifold for three banks of four tanks each. Thus t_{99} refers to the time to expel 99 percent of the mass in a bank of four tanks which contains a volume of 20.55 cubic feet, while t'_{99} refers to the time to expel 99 percent of the mass of all three banks of tanks containing 61.7 cubic feet.

Leaks other than those due to severed tubes or holes of diameter d may occur due to activation of the PRD and to contaminated or damaged fittings. Typical leak rates due to these effects are summarized in the Table B-2.

Table B-1. Initial Leak Rates and Mass Ejection Times

$p_{t,o}$ (psia)	d (in.)	$Q_{std,o}$ (SCFM)	t_{99} (min.)	t'_{99} (min.)
3000	0.75	32,333	*	1.86
	0.50	14,370	1.40	4.20
	0.375	8,083	2.49	7.47
	0.25	3,593	5.60	16.80
	0.125	898	22.4	67.20
1500	0.75	16,166	*	1.86
	0.50	7,185	1.40	4.20
	0.375	4,042	2.49	7.47
	0.25	1,796	5.60	16.80
	0.125	449	22.4	67.20

*There is no tubing this large in the plumbing for the individual banks of four tanks each. This size tube connects the three banks and, therefore, has access to three times the initial mass.

Table B-2. Typical Leak Rates

Leak Type	Q_{std} (SCFM)	Data Source
PRD activation	355	Reference 17
Contamination of fitting	5.8	Reference 11
Scratched ferrule	1.4	Reference 11

REFERENCES

1. E. Pollack, *In the Eye of the Beholder,* Natural Gas Fuels, February 1994.
2. T. Harte, *Alternate Fuels: New York City Sanitation Takes the Plunge . . . Cautiously,* MSW Management, Elements, 1994.
3. R. E. Petsinger, *Capital Metro Station Opens in Austin,* Natural Gas Fuels, January 1994.
4. W. E. Baker et al., *Explosion Hazards and Evaluation,* Elsevier Scientific Publishing, England, 1983.
5. G. A. Karim, *Some Considerations of the Safety of Methane, (CNG), as an Automotive Fuel—Comparison with Gasoline, Propane, and Hydrogen Operation,* Society of Automotive Engineers, paper 830267, 1983.
6. G. A. Karim and I. Wierzba, Safety Measures with the Operation of Engines on Various Alternative Fuels, in *Reliability Engineering and System Safety,* Elsevier Science Publishers, Ltd., England, 1992.
7. A. Booz and Hamilton, Inc., *Alternative Fuels and Facilities Evaluation Study,* prepared for Sacramento Regional Transit District, April 1991.
8. Gage-Babcock and Associates, *Sacramento Regional Transit District, CNG Fueled Bus Maintenance Facility,* study prepared for Pacific Gas and Electric Company, September 1992.
9. T. J. Grant and H. S. Samia, *Safety Analysis of Natural Gas Vehicles Transiting Highway Tunnels,* paper presented at the Intersociety Energy Conversion Engineering Conference, American Society of Mechanical Engineers, 1991; see also Ebasco Services Incorporated Report APTR-46, August 1989.
10. T. J. Grant et al., *Hazard Assessment of Natural Gas Vehicles in Public Parking Garages,* Ebasco Services Incorporated, July 1991.
11. M. J. Murphy et al., *Extent of Indoor Flammable Plumes Resulting from CNG Bus Fuel System Leaks,* paper presented at the Society of Automotive Engineers International Truck and Bus Meeting and Exposition, Toledo, Ohio, November 1992.
12. T. J. Forsythe and R. Kewin, Background on Facilities Modification for Natural Gas Fueled Bus Use: Part 2, *Journal of Applied Fire Science, 5,* pp. 61-75, 1995.
13. B. Gebhart, Y. Jaluria, R. L. Mahajan, and B. Sammakia, *Buoyancy-Induced Flows and Transport,* Hemisphere Publishing Corp., New York, 1988.
14. P. N. Papanicolaou and E. J. List, Investigations of Round Vertical Turbulent Buoyant Jets, *Journal of Fluid Mechanics, 195,* pp. 341-391, 1988.
15. C. J. Chen and W. Rodi, *Vertical Turbulent Buoyant Jets—A Review of Experimental Data,* Pergamon Press, New York, 1980.
16. H. B. Fischer et al., *Mixing in Inland and Coastal Waters,* Academic Press, 1979.
17. *Orion Service Manual, Section 09.5—Natural Gas Fuel System,* p. 09.5.8, February 1993.
18. V. D. Long, Estimation of the Extent of Hazard Areas Round a Vent, in *Second Symposium on Chemical Process Hazards,* Institution of Chemical Engineers, United Kingdom, pp. 6-14, 1963.

CHAPTER 5
Expert Systems Applied to Spacecraft Fire Safety

Richard L. Smith and Takashi Kashiwagi

Expert systems are problem-solving programs that combine a knowledge base and a reasoning mechanism to simulate a human "expert." The development of an expert system to manage fire safety in spacecraft, in particular the NASA Space Station Freedom, is difficult but clearly advantageous in the long term. The report discusses some needs in low-gravity flammability characteristics, ventilating-flow effects, fire detection, fire extinguishment, and decision models, all necessary to establish the knowledge base for an expert system.

This chapter is a general description of expert-system programs, to describe their unique application in meeting the needs of fire safety in spacecraft, in particular the future NASA Space Station Freedom [1-3].

An expert system is a computer program that solves real-world problems whose solution would normally require a human expert [4-7]. Assume you are in communication with a fire-safety expert by the use of a terminal. You type in your questions to him; he in return may ask you questions before he gives you his advice. The quality of his advice is what makes him an expert. If you cannot tell whether you are communicating with a human expert or with a computer running an expert system, the computer program qualifies as an expert system. Normally the domain of discourse must be restricted to a particular field of expertise.

This chapter explains expert systems in terms of their feasibility for development as autonomous intelligence for fire safety in Freedom. A second section of the article discusses the current problems in fire prevention, detection, and control in space, as a preliminary assessment of the presence and absence of expert knowledge in these fields.

In gross terms, an expert system program consists of two major components: a knowledge base and a reasoning mechanism that processes this knowledge to produce additional knowledge. The knowledge base is facts such as the density of iron, the pilot ignition level of dry wood, or the temperature of a circuit board, which are communicated to the knowledge base by a sensor on the circuit board. The reasoning mechanism can be thought of as consisting of two parts: the type of reasoning, either forward or backward reasoning, and the rules that tell what the facts imply. An example may make this clear. Consider the statement, "If the chair is combustible and it has contact with a flame for 5 minutes, then it will start to burn." This is what is called a rule. This rule has two "if" conditions and one "then" action or conclusion item. Thus, if the program has this rule as a forward-reasoning rule and the two "if" conditions are in its knowledge base, it will add to the knowledge base the additional fact that the chair will start burning.

One example of an expert system is MYCIN [8], which diagnoses and recommends treatment for infectious diseases. In comparisons with medical experts in this limited domain of infectious diseases, the performance of MYCIN is shown to be as good as that of the human experts. Another well-known expert system is DENDRAL [9], that uses primarily mass spectrographic and nuclear magnetic resonance data to determine the molecular structure of unknown compounds. DENDRAL's performance is superior to most human experts in its domain. Finally, the expert system that is possibly the greatest commercial success is XCON (or R1 as it was originally called) [10]. It is used by Digital Equipment Corporation to configure computer systems for its customers. Because of the large number of possible combinations of computer components, the problem of getting all the components, cables, etc. together to assemble a working system without missing a part or having items left over is very complex. XCON locates all parts in a reasonable arrangement and plans all the connections. It also verifies that the customer order is correct in that there are no missing nor surplus parts. It has been in constant use since 1980, and its performance has been significantly superior to that of human experts.

The main reason to develop an expert system is to make available to a nonexpert the knowledge and expertise of an expert or experts in the most effective manner. Expert systems have a number of unique advantages. A well designed expert system will give the user the effect of having a consultant on hand. The simulated consultant will be "blind" in that it cannot make direct observations. However, it will have certain advantages over a human consultant. It will always operate at peak efficiency. It will never have an off day. It will be available seven days a week, fifty-two weeks a year. The proposed expert system will have explanation capabilities and infinite patience. It will have the ability to explain all its conclusions and inferences step by step, over and over again. Thus its reasoning will be fully accessible to the program's user. The expert system performance can be made to equal or surpass the performance of most human experts.

EXPERT SYSTEMS FOR SPACE STATION
FREEDOM FIRE SAFETY

Feasibility

The type of expert system best suited for a space station would be an autonomous system with integrated sensing, computing and controlling functions, installed within the station [11]. The system would monitor sensors in the station, to be informed of the current status of the station as it related to fire safety. It would also have the capability to fight the fires or to take any appropriate action in dealing with a fire. The expert system would thus act as an advisor to humans who are either aboard the station or are elsewhere, recognizing potential fire situations and their control. However, in uninhabited space compartments, the system would operate on its own.

In adapting expert systems to meet fire-safety needs in the future U.S. Space Station Freedom, clearly one must deal with some very difficult problems. There are no human experts, recognized as knowledgeable in low-gravity fire conditions, at least comparable to those with expertise in building-fire controls. The development of a fire-safety expert system for Freedom will be slow, costs will be high, and scheduling will be a problem. In addition, the extreme level of required reliability will be an impediment to the development of the system [12].

Nevertheless, Freedom has some unique features that encourage the development of an expert system. As a one-of-a-kind assembly, the physical dimensions of the structure and the materials used will be known. Because of the well-defined structure, one can install smart detectors and a fast extinguishing system in the structure. In addition, there is on-going research on expert-system techniques to provide autonomous operation and maintenance for at least one subsystem in Freedom [11]. The development of a fire-safety expert system can benefit from this research interest and support, and the fire-safety expert system may share existing computer hardware in the space station.

Development Plan

An expert system is normally developed by first using the expertise of only one expert. This facilitates the evaluation of the performance of the computer program and the collection of knowledge. Later, the expertise of other experts can be added to the program.

A suggested method to develop an expert system for Freedom is shown schematically in Figure 1. The flow chart steps are, in brief:

1. Select an expert who will be used as the human expert for the development of the first expert system program.

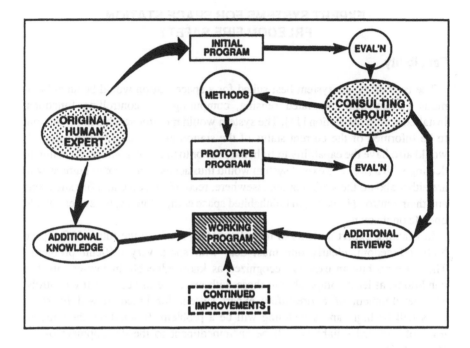

Figure 1. Representation of development of an expert system.

2. Select a small consulting group of human experts who will evaluate the performance of the initial program and provide additional expertise so that the program can be improved.
3. Determine the reasoning methodology of the original expert and produce an initial expert system.
4. From the previous step, identify serious knowledge gaps that should be filled so the program can perform up to expectations.
5. Have a group of experts review the previous two items and develop a consensus for a prototype program.
6. Continue development of the prototype program until its performance is equal to that of the initial expert.
7. Add the knowledge of the group until the program can perform better than any individual member of the group.
8. Continue the development of the working system even after it is deployed, by modifying its knowledge base as new knowledge and operating experience are acquired.

FIRE SCIENCE FOR THE SAFETY OF
SPACE STATION FREEDOM

General Overview

The unique nature of the fire-safety problem in Freedom is discussed in this section. Here we suggest an initial analysis of the fire problem for Freedom and the strategy for dealing with fires in the station. This type of analysis would serve as the starting point for an expert system. In the process of developing an expert system, this analysis would be expanded, refined, and modified.

Information on fire behavior in the low-gravity environment of space is very limited, but it is logical to assume that this behavior is significantly different from that commonly observed in normal gravity [3]. While this lack of basic knowledge complicates the study of spacecraft fire safety, in a sense, the fire-safety problem is well defined. As already mentioned, the fire model for Freedom deals with only a few internal geometric configurations compared to the almost limitless numbers for building configurations. Another important feature is that contents and their location are well specified for Freedom compared to buildings. This allows a detailed, scientific fire-safety study of Freedom to be feasible. Also such a study is desirable for the safety of the crew (no possible immediate evacuation from the station) and is cost effective due to the extremely high program costs of Freedom.

Since there is no possibility of immediate escape of the crew from Freedom, the protection of crew members is the essential element of fire safety in the spacecraft. The second necessity is the preservation of the spacecraft structure if a fire occurs. While these two goals are no different than those of fire safety for buildings in a normal-gravity environment, structural preservation is sometimes compromised in buildings in order to control fires. For this reason, the early detection of fire and its subsequent rapid suppression should be the key points of emphasis for the fire safety of Freedom. The crucial requirement is to keep the fire as small as possible and to extinguish it as soon as possible. Suggested approaches to satisfy these requirements are discussed in the next section. If, unfortunately, the fire cannot be contained in an early stage, a drastic measure to suppress the fire is needed, and this approach is also discussed briefly in the next section.

Early Fire Detection in Space

Various scenarios may be proposed for fires initiated from different sources in Freedom. One such scenario could be a fire initiated from polymeric materials or fluids heated by an accidental local surge in electric power; another could be a fire initiated from an accident in an onboard scientific experiment.

The early detection of a fire caused by an accidental overloading, short circuit, or overheating of an electric device appears to be more difficult in a low-gravity environment than in a normal-gravity environment. The optimum locations of fire detectors are not readily defined in a low-gravity environment due to the lack of a buoyancy-induced upward flow. The flow motion in the spacecraft is determined mainly by a ventilation flow (flow speed of a few centimeters per second) instead of a buoyancy-induced flow. Since the locations and flow rates of vents and the geometrical configuration of the inside of Freedom will be well-specified, calculation of the three-dimensional detailed flow distribution driven by the ventilation inside the modules is feasible, using commercially available engineering codes as a first step [13]. Besides this overall flow calculation, a flow-pattern calculation inside various electronic cabinets, each with its own ventilation, is needed to be able to predict flow patterns of early fire products. These flow patterns can be used to predict two important aspects of fire behavior. One is the spatial distribution of fire products in the spacecraft using data on the generation rates of these products; the other is the possible fire growth in an electric cabinet and also in the general space-station volumes using the information on material flammability characteristics. The optimum locations of detectors for early fire detection can be strategically identified from the flow patterns and potential high-risk fire areas. Furthermore, it should be possible to determine the location of fire from the observed pattern of activated detector locations. The complexity of these calculations and the need for real-time response dictate the storing of a library of precomputed and validated results.

The selection of the type of detector sensors is crucial not only to detect fire as early as possible but also to avoid false alarms. The use of a combination of different types of sensitive detectors at various locations might provide useful data as an input to an alarm/no alarm algorithm (rules, decision). The algorithm would contain information from the calculated flow patterns both in the electronic cabinets and in the general spacecraft volumes and it could make a decision whether the detected "fire" is real or false based on the flow pattern applied to the location and type of the activated detectors.

The use of different types of detectors, spatially distributed detectors, and a decision algorithm is aimed at reducing false alarms as much as possible without sacrificing the capability of early fire detection. However, information on fire products is needed for the selection of detector sensors. For example, it is known that an ionization detector is more sensitive to small particulates (< 0.3 μm) but a light-scattering detector is more sensitive to large particles (> 0.3 μm). It is not clear what sizes of particulates are formed from a fire in a low-gravity environment. Smoldering, pyrolysis, and localized flaming are possible in the early phase of a fire. In a normal-gravity environment, particulates generated from smoldering and pyrolysis tend to be condensed liquids and to be larger than those generated from flaming of the same material. An electrochemical CO detector and a flame-ionization detector can be used in combination with a particle-sensitive detector. If quantitative characterization of the generation of the products from smoldering, pyrolysis, and small flames of polymeric materials appropriate for spacecraft is made, the concentration and size distribution of particulates and also concentrations of CO and other gases can be calculated at various positions in the spacecraft by combining the material-products characteristics with the above flow calculation. The products from smoldering and pyrolysis may not be significantly affected by gravity, but those from small flames might be. Extensive studies are needed to obtain the relationship between the fire products in a low-gravity environment and those in a normal-gravity environment. The calculated concentration distribution in both electronic cabinets and in the general spacecraft volumes will provide detailed information for the selection of the strategic locations of detectors and also for threshold criteria of toxic hazards for the crew. This information can also be used for the determination of the best selection of the crew's evacuation route and method (stay near the floor or float near the ceiling).

The screening of polymeric materials which are going to be used in Freedom can contribute significantly to the fire safety of Freedom. Although it is generally considered that polymeric materials are less flammable in a low-gravity environment than in a normal-gravity environment (there are limited experimental data [14, 15]), more studies are needed to obtain quantitative relationships between the two different gravity environments. As discussed above, knowledge of the characteristics of smoldering, pyrolysis, ignition, and flame spread of the polymeric materials is important in order to predict fire behavior. Thus, low-gravity flammability tests should be used not only for screening of the materials but also to measure quantities useful for predicting fire scenarios. The measured material-flammability characteristics should be stored in the algorithm for early fire detection and fire growth analysis, as described above. Products distribution (particulates, CO, CO2, total hydrocarbons, acutely toxic gases such as HCN, HCL and acrolein) and their generation rates should be measured under smoldering, pyrolyzing, and small flaming modes.

Fire Suppression in Space

After the detection of a fire, rapid suppression of the fire is essential. As discussed above, an early fire may be in one of the possible modes of smoldering, pyrolyzing, or small flaming for polymeric materials (cables, insulation materials, papers and so on) or fluids. The suppressant must suppress (extinguish) the three types of fire modes. The present Shuttle extinguishing agent, a halogenated hydrocarbon or halon [1, 3], is recognized as an excellent extinguisher for the flaming-fire mode, but it is less effective for the other two types. Furthermore, the use of halons raises questions of the toxic and corrosive nature of their reaction products in flames. A fine water spray is effective for the extinguishment of the flaming-fire mode, but its effectiveness for the other two modes in a low-gravity environment is questionable due to the low sticking efficiency of water droplets onto a smoldering or pyrolyzing surface. Furthermore, the moisture concentration in the atmosphere could become higher with the use of water, which tends to increase corrosivity of metals by fire products, and cleaning of this excess water after the fire might be a problem.

A promising concept for a fire suppressant for space is a fire-extinguishment foam generated with compressed nitrogen. The foam should have a good sticking nature on the surface in a low-gravity environment (however, apparently little is known about the performance of foams in low gravity), and also it is an effective extinguishing agent for all three fire modes. Furthermore, the total mass of liquid is so small that cleaning after the fire is relatively simple. The foam must have a very high dielectric constant to avoid any shorts in electronic devices. Many small cans of the foam extinguishers (like a shaving cream) can be strategically installed at potential fire-hazard areas for use as portable fire extinguishers or as fixed extinguishers to be actuated remotely to fill up the designated compartment. Information regarding the relationship between the amount of the foam and an extinguishable fire size would be needed to test the plausibility of this approach. Since fire behavior at low gravity is poorly understood, its suppression requires imaginative and thorough development.

Drastic Measures

If the above fire-fighting tactics fail and a fire becomes large enough to threaten the safety of the crew or the integrity of the spacecraft, some drastic measure is needed to extinguish the fire as quickly as possible. The best approach appears to be venting the module to the outside vacuum of space as described below. Here it is assumed that the module is compartmentalized and can be vacuum tight. It is not necessary to reduce the pressure to vacuum levels but only to a low total pressure sufficient to ensure that the remaining oxygen quantity (partial pressure) in the environment cannot sustain any fire. While inert-gas pressurization has been investigated for extinguishment of difficult fires in confined compartments in submarines [16], venting to space is much faster than inerting and does not require a large quantity of inert gas storage. Also, when venting is employed, there is no need to provide the extra structural weight required to deal with the problem of over-pressurization of the spacecraft structure. Halons could be also used for this stage of a fire, but there might be some serious reservations due to possible toxic and corrosive natures of halon-flame reacted compounds. Although venting might increase or transfer fire momentarily along the flow to the venting opening [3], the rapid decrease in pressure in the module should suppress the fire quickly. Another advantage of venting is the rapid dilution of harmful by-products in the atmosphere; hence, halons or water may be used for control of difficult fires followed by venting to avoid the deleterious effects of these extinguishants on the atmosphere. One disadvantage of venting is that the vented products may coat the outside surface of windows of the spacecraft with deposits; this might hinder certain scientific experiments. This problem has been experienced with normal waste-product venting and micrometeorite etching in the Soviet space-station program [17]. Potential solutions include retractable transparent covers. Obviously, the consequences and benefits of suppression by venting to vacuum needs a thorough analysis.

Another important decision is under what conditions drastic venting is to be applied. The decision should be based on the safety of the crew and the structural integrity of the space station. CO concentrations, temperatures, pressure, and other properties that relate to the safety and the structural integrity should be considered as indicators to constitute a decision algorithm regarding venting. This algorithm should include possible fire scenarios and calculated potential fire-growth histories based on fuel loading and material flammability characteristics. The approach to a lethal CO concentration level or maximum temperature and pressure levels should be used as one criterion for the venting decision in the algorithm. Therefore, the algorithm can be one large code that includes the results of flow-pattern calculations in electronic cabinets and in the general spacecraft volumes, the decision analysis on early fire detection, the collection of CO concentrations, temperatures and other properties, and finally the decision analysis on the actuation of venting.

CONCLUDING REMARKS

This report explores the capabilities of expert-system programs in their application to fire-safety management in future spacecraft, a new role for these programs. The concept and development of expert systems are briefly described. The goal of this review is the assessment of expert systems for autonomous and crew-assisted fire-safety management in the NASA Space Station Freedom, now under design.

Although the development of an expert system for spacecraft fire safety can be very costly and time-consuming, this application has clear advantages in safety operations. What is lacking at present is an adequate knowledge base to simulate that of a human "expert." This report thus includes a discussion of fire-safety problems in spacecraft. From these findings, it is possible to identify some important elements of new knowledge or capabilities needed to establish an expert-system program for Freedom. These items are, in summary:

1. Determination of flammability characteristics of combustible materials in the low-gravity environment of Freedom

2. Development and validation of the ventilating and exhaust-flow calculations in electronic cabinets and in the general module volumes
3. Determination of the generation rate and characteristics of fire products from smoldering, pyrolysis, and small flaming of polymeric materials in a low-gravity environment
4. Development of new detectors, such as a highly selective CO detector
5. Development of new fire extinguishers, such as foams, and determination of their properties and extinguishment characteristics
6. Development of an algorithm or algorithms to make the decision as to a false alarm or a real fire from the collection of CO concentrations, temperatures, and other properties, and to make the decision of the venting based on the stored possible fire-growth histories

Like a person entering a new field or discipline, a program on its way to becoming an expert program will start as a novice. If it continues its development, it eventually becomes an expert. Just as an apprentice can help the journeyman, an apprentice expert system will be of significant value, and a journeyman expert system will be of even greater value. Thus we see the evolution of a series of expert systems that will be of great value in their domain of interest to Space Station Freedom. Each of these programs will improve the quality of the decision-making of the human expert that it aids.

REFERENCES

1. *Spacecraft Fire Safety,* NASA CP-2476, 1987.
2. G. A. Rodney, *Safety Considerations in the Design of Manned Spaceflight Hardware,* IAF Paper 87-569, October 1987.
3. R. Friedman and K. R. Sacksteder, *Fire Behavior and Risk Analysis in Spacecraft,* NASA TM-100944, 1988.
4. R. L. Smith, *ASKBUDJR: A Primitive Expert System for the Evaluation of the Fire Hazard of a Room,* NBSIR-86/3319, March 1986.
5. S. M. Weiss and C. A. Kulikowski, *A Practical Guide to Designing Expert Systems,* Rowman & Littlefield Publishers, Totowa, New Jersey, 1984.
6. A. Goodall, *The Guide to Expert Systems,* Learned Information, Inc., Medford, New Jersey, 1985.
7. P. H. Winston and K. A. Prendergast (eds.), *The AI Business: Commercial Uses of Artificial Intelligence,* MIT Press, Cambridge, Massachusetts, 1984.
8. B. Buchanan and E. Shortliffe, *Rule Based Expert Systems: The MYCIN Experiments of the Stanford Heuristic Programming Project,* Addison-Wesley Publishers, Reading, Massachusetts, 1984.
9. R. K. Lindsay et al., *Applications of Artificial Intelligence for Organic Chemistry: The DENDRAL Project,* McGraw-Hill Book Co., New York, 1980.
10. J. Bachant and J. McDermott, *R1 Revisited: Four Years in the Trenches, AI Magazine,* 5:3, pp. 21-32, Fall 1984.

11. H. Lum and E. Heer, *Progress toward Autonomous, Intelligent Systems,* IAF Paper 87-31, October 1987.
12. E. Heer and H. Lum, Raising the AIQ of the Space Station, *Aerospace America,* 25:1, pp. 16-17, January 1987.
13. R. G. Davis and J. L. Reuter, *Intermodule Ventilation Studies for the Space Station,* SAE Paper 871428, July 1987.
14. C. R. Andracchio and T. H. Cochran, *Gravity Effects on Flame Spreading over Solid Surfaces,* NASA TN D-8228, 1976.
15. S. L. Olson, *The Effect of Microgravity on Flame Spread over a Thin Fuel,* NASA TM-100195, 1987.
16. P. A. Tatem, R. G. Gann, and H. W. Carhart, Pressurization with Nitrogen as an Extinguishant for Fires in Confined Spaces, *Combustion Science and Technology,* 7:5, pp. 213-218, 1973.
17. K. P. Feoktistov and E. K. Demchenko, Description, Photos of Salyut-7 Station, *USSR Report: Space,* No. 20, JPRS 82970; Joint Publishing Research Service, Arlington, Virginia, pp. 8-16, February 1983 (translation from *Zemlya i Vselennaia,* No. 6, pp. 11-16, November-December 1982).

CHAPTER 6
Assessing Community Fire Risk: A Decision Analysis Based Approach

Abel A. Fernandez, Derya A. Jacobs,
Paul R. Kauffman, Charles B. Keating,
and Dennis C. Sizemore

A prototype community fire risk assessment study was performed for accreditation under the National Fire Service Accreditation Program, developed by the International Association of Fire Chiefs (IAFC). The fire risk model used for the study quantifies risk as a function of perceived consequence from a fire and probability of fire at a structure. The consequence from a fire is determined through expert judgment using the Analytic Hierarchy Process (AHP). The probability of fire at a structure is derived from empirical evidence. A simple additive weighting scheme is employed to combine the consequence and probability factors into a single valued measure of perceived fire risk potential for a structure. A classification scheme is then used to group structures according to risk level. The results of this study provide a model and risk assessment tool for fire science practitioners.

MOTIVATION FOR THE STUDY

The National Fire Service Accreditation Program developed by the International Association of Fire Chiefs (IAFC) and the International City Management Association (ICMA) provides a voluntary means through which fire departments can assess when they have achieved an appropriate level of organizational performance and efficiency. Through a self-assessment process, the individual fire departments examine the adequacy of a community's standards of coverage (i.e., their policies and procedures) based on a defined set of metrics. Certification hinges on the establishment of policies and procedures that ensure achievement of community goals, objectives and mission.

The Fire and Emergency Service Self-Assessment Manual published by the IAFC National Fire Service Accreditation Task Force provides general process guidance and requirements for accreditation [1]. One of the requirements for certification is that the fire department perform a community fire risk assessment. A challenge in pursuing certification is that the guidelines and requirements for accreditation do not detail how to conduct the self-assessment process. Furthermore, the literature is wanting in providing a methodology for effectively and efficiently conducting a self-assessment. This project proposes an analytical approach for conducting community fire risk assessments. The objective of the project was to develop a framework that fire departments could easily adapt to meet their needs in conducting a community fire risk assessment. The study developed and implemented a conceptual fire risk model for one of the fire districts within the City of Hampton, a community of approximately 130,000 people in southeastern Virginia. The model quantifies risk as a function of consequence and probability of occurrence. The consequence of a fire at a particular structure is determined through expert judgment using the Analytic Hierarchy Process (AHP). The probability of fire at a structure is derived from empirical evidence. Once the risk at a structure is quantified, a classification scheme is used to group structures according to risk level. These risk groupings provide the means to construct a community fire risk profile and serve as potential input for other studies such as fire district simulation models. In addition, the risk profile provides valuable information for strategic decision making by local fire safety officials.

A FIRE RISK MODEL

In accordance with the Fire and Emergency Service Self-Assessment Manual, the objective of a community risk assessment study is to classify each structure within a fire district according to risk potential. The study reported in this article adopted a two-dimensional risk model that measures risk according to consequence and probability, as shown in Figure 1 [2]. Consequence is a measure of the impact from a fire, while probability reflects the relative frequency of fire occurrence. The fire risk of a particular structure is quantified by a (Consequence, Probability) doublet, in contrast to a more traditional approach that defines risk as the product of consequence and probability (an expected value of consequence) [3]. Under the latter concept, risk is seen as an expected value of loss. It may thus be possible for two structures to have the same risk (expected value of consequence); yet one structure may have a high probability and low consequence and the other a low probability and high consequence. Maintaining consequence and probability as separate measures recognizes that assessing fire potential risk requires an understanding of the potential impacts resulting from a fire *and* the associated likelihood of fire occurrence.

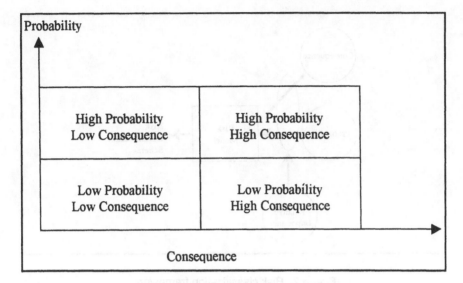

Figure 1. A fire risk model.

The fire risk model illustrated in Figure 1 provides the framework for classifying structures according to risk potential. This framework requires three separate inputs to determine the level of risk potential for each structure in the community, as shown in Figure 2. The first input required is the consequence from a fire at a particular structure. The community must assess the consequences and their associated impact from a fire at the structure. Determining the impact from a fire requires subjective assessment of the degree of consequence severity associated with fires at the different structures in the district. This study uses the Analytical Hierarchy Process (AHP) as a rational, disciplined approach for eliciting and quantifying expert opinion regarding the consequences from fires. The second input required is probability of fire occurrence. Here quantification relied on statistical analysis of empirical data to determine the relative probability of fire occurrence for particular types of structures. Lastly, the third input required is subjective judgment concerning the combination of consequence and probability to assign structures to risk classes. This requires assessment of the relative importance between consequence and probability. A very high consequence and very high probability structure is naturally classified as having a maximum risk potential. Similarly, a very low consequence and very low probability structure is naturally classified a minimum risk potential. Assigning risk classifications becomes complex for those structures not exhibiting extreme consequence and probability characteristics. In those instances, the appropriate risk classification hinges on the community's perception of risk. A particular community may view a low probability, but high consequence fire, as catastrophic and thus classify

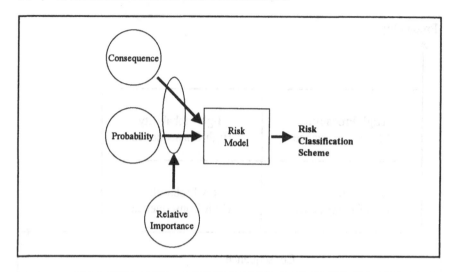

Figure 2. Risk classification framework.

a structure with those characteristics as a maximum risk. The approach taken accommodates the perceptions and context of the community when assigning risk classifications.

The objective of the community risk assessment study is to classify each structure within a fire district according to a defined measure of risk. In a large community such as Hampton, the number of structures makes individual classification on a structure by structure basis impractical. The following section describes the approach used to group structures so as to facilitate the risk assessment process.

Grouping Structures Within the Fire District

Accreditation requires that every structure within a community be classified according to risk level. Previous studies suggested the use of Building Officials Code Administrators (BOCA) codes as a useful mechanism for assigning risk to structures [4-7]. BOCA codes classify structures according to type of occupancy, and are particularly convenient for present purposes since structures, almost without exception, are classified according to the BOCA codes. These codes are a convenient vehicle for a risk classification scheme as they:

1. Are accepted within the fire protection community,
2. Are readily available for every structure, and
3. Classify structures according to type of occupancy (a scheme appropriate for the present purposes).

The BOCA codes used in the risk assessment study are presented in Table 1. A complete description of each classification is provided in Chapter 3 of the Building Officials Code Administrators National Building Code [7].

The BOCA codes shown in Table 1 provide a convenient grouping for assigning consequence and probability to each structure within a community. The values of consequence and probability associated from a fire are assessed on a BOCA code basis (a homogenous cluster of similar structures) rather than on an individual structure basis. This approach greatly reduces the dimensions of the problem from the thousands of structures within the fire district to twenty-four, the number of principal structural classifications within the BOCA code. The following sections describe the approach used to assign values of consequence and probability to each BOCA code.

AHP AS A TOOL FOR DEVELOPING FIRE CONSEQUENCE VALUES

Using BOCA codes as a convenient means of grouping structures within a community, the first input needed for the fire risk model is the severity of consequence if a fire should occur. This assessment requires that the community identify the factors affecting the degree of impact from a fire at a given type of structure. Furthermore, the impact from the various factors must be consolidated through some weighting scheme to derive a single value representing the consequence from a fire at a particular type of structure. This type of problem is classified as a multiple criteria decision-making problem, as it requires the analysis of multiple factors impacting a decision [8]. Among available multi-criteria decision analysis methods, the AHP was found to be best suited for this application.

The AHP is a well accepted tool for multiple criteria decision making, as is evident from the vast literature of research and application papers [9, 10]. In this particular application, AHP provides a process which:

1. Structures the problem using an easily understood model showing the inter-relationship between the various factors;
2. Elicits expert opinion through a theoretically sound procedure that represents subjective judgments with meaningful metrics; and,
3. Develops normative standards that may be used to rate the severity of the consequences associated with a fire.

An important characteristic of the AHP is that it provides a coherent framework for group discussion and debate leading to consensus. As discussed in the following section, the AHP involves the decision makers in all phases of model building and judgment processes. This personal involvement, in easily understood terms, fosters decision maker buy-in to both the model and the ultimate results.

Table 1. Summary Descriptions of BOCA Codes

Code	Summary Description
Assembly	**Designed for gathering of people**
A1	Stage and fixed seats
A2	Without stage
A3	Without fixed seats
A4	Place of worship
A5	Outdoor assembly
Business	**Occupied for transaction of business**
B	Business
Factory and Industrial	**Occupants engaged in product labor**
F1	Moderate hazard
F2	Low hazard
High Hazard	**Associated with hazardous materials**
H1	Detonation hazard
H2	Deflagration hazard
H3	Physical hazard
H4	Health
Institutional	**Occupants have limited mobility**
I1	Elderly homes
I2	Hospitals
I3	Prisons/jails
Residential	**Sleeping accommodations excluding institutions**
R1	Hotels
R2	Multiple family
R3	One/two family
R4	Detached dwelling
Storage	**Storage of goods or merchandise**
S1	Moderate hazard
S2	Low hazard
Educational	**Non-business educational structures**
E	Educational purposes
Mercantile	**Occupied for display and sales purposes**
M	Retail sales structures
Utility & Miscellaneous	**Structures not classified elsewhere**
U	General structures not classified elsewhere

THE AHP MODEL

The AHP provides a framework for modeling the factors impacting the consequence from a fire, eliciting expert opinion regarding the relative importance of each factor, and translating these into priority weights for assessing the total consequence. This procedure ensures that well established norms are used as the basis for determining the consequence associated with each BOCA code structural type. The keystone of the AHP is a hierarchical model depicting at the first level the principal factors (attributes) impacting the degree of consequence severity, and at a second level the intensities of each factor. Figure 3 presents the AHP model developed for quantifying the consequence severity associated with a fire.

The model shown in Figure 3 follows the usual AHP format. At the top of the hierarchy, Level 0 in AHP terminology, is the principal objective or question addressed by the model. In this case, our objective is to determine the overall severity of the consequence from a fire. The next level in the hierarchy, Level 1, identifies the principal factors impacting the severity of the consequence (the level immediately above). As shown in Figure 3, three factors are identified as having the most significant impact upon the overall severity of the consequence:

1. The concentration of occupants within the structure;
2. The presence of hazardous materials within the structure; and
3. The mobility of the occupants within the structure.

The bottom-most level of the AHP model, Level 2, represents the possible intensity levels for each of the factors at the second level. These intensities break down each factor into mutually exclusive and exhaustive classifications for the strength of presence of each factor, i.e., the intensity with which the factor is present at a

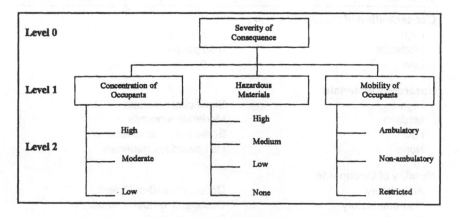

Figure 3. AHP model for consequence.

particular structure. Thus, for example, the factor Concentration of Occupants within a structure is broken down into the three intensity levels of High, Moderate or Low. The individual intensities are clearly defined so as to minimize the possibility of classification error. Table 2 summarizes the definition of each intensity.

The AHP model shown in Figure 3 represents the principal factors impacting the severity of the consequence from a fire at a particular structure. The next step in the AHP process is to elicit expert opinion concerning the relative importance of each of these factors. These judgments form the basis for deriving norms by which each BOCA code structural type is assigned a quantitative measure representing the consequence from a fire.

Determining Relative Importance through Expert Elicitation

A quantitative assessment of the degree of consequence severity is needed in order to assess the risk potential of a particular structure. The previously described AHP model provides a means by which an objective function expressing the consequence of a fire can be expressed as a weighted linear function of the three attributes (level two factors) of the hierarchy. This object function takes the following form:

Consequence = w1 (Concentration) + w2 (HazMat) + w3 Mobility) where,
Consequence ≡ quantitative measure of the consequence from a fire;
Concentration ≡ quantitative variable reflecting the concentration of occupants;
HazMat ≡ quantitative variable reflecting presence of hazardous materials;

Table 2. Summary of Factor Intensities

Concentration of Occupants

High	≥ 50 people
Moderate	6-50 people
Low	< 6 people

Hazardous Materials

High	Significant amounts
Medium	Moderate amounts
Low	Some but minimal amounts
None	No hazardous materials

Mobility of Occupants

Ambulatory	Occupants self-sufficient
Non-ambulatory	Occupants require assistance
Restricted	Occupant mobility must be granted

Mobility ≡ quantitative variable reflecting the mobility of occupants; and,
Wj ≡ weighting factor, for j = 1, 2, 3.

Following the usual AHP procedure, the priority weights w1, w2, and w3 are derived from expert judgment using pairwise comparisons of the importance (i.e., dominance) between two elements of the hierarchy relative to the parent element in the adjacent upper level. As shown in Table 3, a 1 through 9 numerical scale is used to quantify the degree of dominance between the two elements. The reciprocal of the values shown in Table 3 are used to designate that one element is less dominant than a second.

For example, with respect to the three possible intensities for the Concentration of Occupants factor, three pairwise comparisons are made:

1. The relative dominance of High to Moderate;
2. The relative dominance of High to Low; and,
3. The relative dominance of Moderate to Low.

Following the same procedure a total of fifteen pairwise comparisons are needed to completely define all of the pairwise relationships between the elements in the AHP model shown in Figure 3. The AHP software package Expert Choice was used to elicit and consolidate the opinion of appropriate experts from the fire district [11]. This software incorporates a number of features to facilitate the expert opinion elicitation process. For example, judgments concerning the relative importance between two elements may be performed using a number of alternative modes including numerical (such as that illustrated in Table 3), verbal, matrix, questionnaire, and graphical. Appropriate diagnostics within the software guide the analyst and experts toward the mode best suited for a particular application. Table 4 presents the consensus opinion with respect to the three intensities for the Concentration of Occupants factor. As shown, the intensity level High is judged to be extremely more important than the intensity level Low. All other pairwise

Table 3. Pairwise Comparison Scale

Intensity of Importance	Definition	Explanation
1	Equal importance	Two criteria contribute equally.
3	Moderate importance	Slightly favor one element over the other.
5	Strong importance	Strongly favor one element over the other.
7	Very strong importance	Strongly favor one element over the other, its dominance is demonstrated in practice.
9	Extreme importance	Evidence favoring one element over the other is of the highest possible order of affirmation.
2, 4, 6, 8	For compromise between above values	

Table 4. Pairwise Comparison Matrix—Concentration Factor

	High	Moderate	Low
High	1	5	9
Moderate	1/5	1	3
Low	1/9	1/3	1

comparison judgments are shown in Table 4. Note that all diagonal elements of the comparison matrix are obviously set to 1 and that diagonally symmetric values are reciprocals.

In addition, the AHP provides a consistency ratio, typically termed CR, which quantifies the inconsistency among the set of all expert judgments [9]. The value of CR is in the range [0, 1], and it provides a measure of rational consistency among all pairwise comparisons of a set. CR values less than 0.10 are considered acceptable. The CR values obtained from the experts at the fire district ranged from 0 to 0.09, and were thus well within the acceptable limits.

The Expert Choice software performs the AHP calculations necessary to determine the values for the priority weights w1, w2, and w3. Based on the expert assessments, the final form of the objective function is:

$$\text{Consequence} = w1 \, (\text{Concentration}) + w2 \, (\text{HazMat}) + w3 \, (\text{Mobility})$$
$$= 0.636 \, (\text{Concentration}) + 0.287 \, (\text{HazMat}) + 0.078 \, (\text{Mobility})$$

These results imply that the factor Concentration of Occupants is judged to be the overwhelmingly most important determinant of Consequence. Note that similar functional relationships are derived between the three level 2 factors and their respective intensity levels. These derived relationships are:

$$\text{Concentration} = w1,C \, (\text{High}) + w2,C \, (\text{Moderate}) + w3,C \, (\text{Low})$$
$$= 1.0 \, (\text{High}) + 0.237 \, (\text{Moderate}) + 0.094 \, (\text{Low})$$

$$\text{HazMat} = w1,H \, (\text{High}) + w2,H \, (\text{Medium}) + w3,H \, (\text{Low}) + w4,H \, (\text{None})$$
$$= 1.0 \, (\text{High}) + 0.550 \, (\text{Medium}) + 0.306 \, (\text{Low}) + 0.082 \, (\text{None})$$

$$\text{Mobility} = w1,M \, (\text{Ambulatory}) + w_{2,M} \, (\text{Non-ambulatory}) + w_{3,M}$$
(Restricted)
$$= 0.111 \, (\text{Ambulatory}) + 1.0 \, (\text{Non-ambulatory}) + 1.0 \, (\text{Restricted})$$

where,

$wi,j \equiv i^{th}$ weighting factor and j = C, H and M.

Note that the intensity variables are all (0, 1) variables, where 1 indicates the presence of that intensity. For any particular structure, only one of the intensity variables for a given factor may take on the value of 1.0.

The above relationships can be used to derive the perceived consequence for a particular structural type. For example, the value of Consequence for a structure with a Concentration intensity of Moderate, a HazMat intensity of Medium, and a Mobility intensity of Ambulatory is calculated as:

Concentration = 1.0 (High + 0.237 (Moderate) + 0.094 (Low)
$\qquad$ = 1.0 (0) + 0.237 (1.0) + 0.094 (0)
$\qquad$ = 0.237
HazMat $\quad$ = 1.0 (High) + 0.550 (Medium) + 0.306 (Low) + 0.082 (None)
$\qquad$ = 1.0 (0) + 0.550 (1.0) + 0.306 (0) + 0.082 (0)
$\qquad$ = 0.550
Mobility $\quad$ = 0.111 (Ambulatory + 1.0 (Non-ambulatory) + 1.0 (Restricted)
$\qquad$ = 0.111 (1.0) + 1.0 (0) + 1.0 (0)
$\qquad$ = 0.111
Consequence = 0.636 (Concentration) + 0.287 (HazMat) + 0.078 (Mobility)
$\qquad$ = 0.636 (0.237) + 0.287 (0.550) + 0.078 (0.111)
$\qquad$ = 0.317

Determining the Consequence for Each BOCA Code

The functional relationships developed through the AHP are used to quantify the degree of consequence severity for each of the BOCA codes. Fire district expert judgment, through consensus, is used to assign intensity levels for each of the twenty-four structural groupings of the BOCA codes. The Expert Choice software facilitates this process by providing a user friendly human interface and performing all necessary calculations to determine the Consequence value for each code. These results are summarized in Table 5.

As shown in Table 5, the derived values of consequence range from 0.731 for structures designated as A1, A2, A3, A4, and R1 (all structures with a High intensity factor for Concentration of Occupants) to 0.092 for structures designated as U (structures used as utility buildings or otherwise not designated elsewhere). The values of consequence are unitless and should be interpreted on a ratio scale basis. That is, the perceived consequence from a fire at a structure with a value of 0.50 is twice that of a structure having a consequence value of 0.25.

The intensity designations shown in Table 5 may be fine tuned to more accurately reflect the characteristics of specific structures within the fire district. For example, the F1 and F2 codes (structures used as factories) are given a Moderate intensity for the Concentration of Occupants factor, reflecting the fact that most factories in the fire district employ between six and fifty people. Specific factories known to employ more than fifty people may be assigned a High intensity, with their consequence value modified to reflect this characteristic.

Table 5. Summary of Derived Consequence Values

BOCA	Consequence	Concentration of Occupants	Hazardous Materials	Mobility of Occupants
A1	0.731	High	Low	Ambulatory
A2	0.731	High	Low	Ambulatory
A3	0.731	High	Low	Ambulatory
A4	0.731	High	Low	Ambulatory
R1	0.731	High	Low	Ambulatory
H1	0.446	Moderate	High	Ambulatory
H2	0.446	Moderate	High	Ambulatory
H3	0.446	Moderate	High	Ambulatory
H4	0.446	Moderate	High	Ambulatory
M	0.317	Moderate	Medium	Ambulatory
F1	0.317	Moderate	Medium	Ambulatory
S1	0.317	Moderate	Medium	Ambulatory
I2	0.316	Moderate	Low	Non-ambulatory
I3	0.316	Moderate	Low	Restricted
B	0.247	Moderate	Low	Ambulatory
I1	0.247	Moderate	Low	Ambulatory
F2	0.247	Moderate	Low	Ambulatory
E	0.247	Moderate	Low	Ambulatory
S2	0.247	Moderate	Low	Ambulatory
R2	0.247	Moderate	Low	Ambulatory
R3	0.156	Low	Low	Ambulatory
R4	0.156	Low	Low	Ambulatory
U	0.092	Low	None	Ambulatory

The AHP model provides a means to derive a value for the perceived consequence from a fire at the various types of structures found within the fire district. Both technical and non-technical personnel easily understand the process used, with key fire personnel providing the expertise for building the model and its associated parameters. The next step in the risk assessment is to develop the probability of fire for each of the BOCA code structures found within the fire district.

PROBABILITY OF FIRE

Recall that the doublet (Consequence, Probability) defines the perceived fire risk associated with a particular structure. The probability of fire at a particular type of structure during a given year is derived from statistical data, in this instance fire data from calendar year 1997. Table 6 summarizes the number of fire calls received by the fire district during a nine month period of calendar year

Table 6. Probability of Fire by BOCA Code

BOCA	No. of Structures	No. of Fire Calls	Calls/ Structures	Probability
A1	0	0	NA	0
A2	16	2	0.125	0.033
A3	1	0	0	0
A4	30	8	0.267	0.070
B	34	20	0.588	0.155
F1	0	0	NA	0
F2	3	0	0	0
H1	0	0	NA	0
H2	0	0	NA	0
H3	9	0	0	0
H4	0	0	NA	0
I1	1	0	0	0
I2	7	13	1.857	0.491
I3	0	0	NA	0
R1	15	0	0	0
R2	7723	349	0.045	0.012
R3	65	32	0.492	0.130
R4	0	0	NA	0
S1	70	0	0	0
S2	0	0	NA	0
E	8	0	0	0
U	0	0	NA	0
M	73	30	0.411	0.109

1997, segregated according to BOCA code. This study period was determined strictly on the basis of available data, and for present purposes the assumption is made that each fire call represents the occurrence of a fire. The second and third columns, respectively, of Table 6 show the number of structures and fire calls for each BOCA code type. Thus, there were two fire calls to the sixteen buildings of type A1 over the nine month period. In order to compensate for the structural mix found in the fire district, the number of fire calls are standardized to calls per individual structure of each particular BOCA code type. These figures are shown in the fourth column of Table 6. For example, over the period of study each A1 structure in the fire district made the equivalent of 0.125 fire calls.

The figures given in the fourth column of Table 6 are frequency of fire for each specific structure within the fire district. These values may be used to derive the required probability of fire expressed as a conditional probability. That is, when a fire call is received what is the probability that the fire is in a specific structure within the fire district? For each BOCA code type, this conditional probability is found by dividing the number of calls per structure by the sum total number of calls per structure over all BOCA code types.

$$P_i = \frac{F_i}{\sum_j F_j} \tag{1}$$

where,

$P_i \equiv$ probability of fire at a particular structure of type i, for i = A1, A2, . . . , M; and,
$F_i \equiv$ fire calls per structure of type i, for i = A1, A2, . . . , M.

The probability for each of the BOCA structural types is shown in the fifth column of Table 6. Thus, for example, given that a fire call is received, the probability that it comes from *a specific* A2 structure is 0.033.

$$P_{A2} = \frac{F_{A2}}{\sum_j F_j}$$

$$= \frac{0.125}{3.786}$$

$$= 0.033$$

Not surprisingly, even though the vast preponderance of fire calls come from multiple family structures (type R2), the probability that a call comes from one specific structure of type R2 within the fire district (from the set of the 7,723 R2 structures in the fire district) is small.

FIRE RISK POTENTIAL OF STRUCTURES

Scaling of Results

Previous sections describe the derivation of values for the perceived consequence from a fire at a particular structure and the probability of fire occurrence, Tables 5 and 6 respectively. Together, the doublet (Consequence, Probability) represents the fire risk potential for each structural type. The intent of the community fire risk study is to assess the fire risk potential of each structure *relative* to the other structures within the community. In order to provide relative values, the consequence and probability measures are scaled using

$$R_{i,j} = \frac{X_{i,j}}{X_{m,j}} \tag{2}$$

where,

$R_{i,j} \equiv$ scaled attribute, for i = A1, A2, ... , M, and j = Consequence or Probability;

$X_{i,j} \equiv$ attribute to be scaled, for i = A1, A2, ... , M, and j = Consequence or Probability; and,

$X_{m,j} \equiv$ maximum value of attribute to be scaled, where m corresponds to the ith index of Max $(X_{i,j})$ and j = Consequence or Probability.

This procedure converts consequence and probability to scaled values in the range [0, 1], where a 0 represents no consequence (or probability) and a 1.0 represents the largest value in the set of consequence values. All $R_{i,j}$ are unitless and represent a structure's consequence (or probability) relative to those of other structures in the community. The second and third columns of Table 7, Summary of Risk Values, contain scaled values of consequence and probability for each of the BOCA codes. For example, the scaled value of consequence for I2 structures is 0.43, derived as

$$R_{I2,C} = \frac{X_{I2,C}}{X_{A4,C}}$$

$$= \frac{0.316}{0.731}$$

$$= 0.43$$

Similarly, the scaled value of probability for the I2 structures is 1.00. Note that structural types not found within the fire district are assigned a fire probability of 0.

Derivation of a Single Valued Risk Metric

According to the risk model shown in Figure 1, the scaled values of consequence and probability can be plotted to visually represent the fire risk potential of the different structural types found in the fire district. The resultant fire risk potential model for the fire district is shown in Figure 4. As previously discussed, structures with a (Consequence, Probability) value of (1.0, 1.0) clearly have the highest possible risk while those with a value of (0, 0) have the lowest risk. The arrow in Figure 4 indicates direction of increasing risk, so that risk increases as one moves from the origin to the point (1.0, 1.0).

The fire risk model shown in Figure 4 only allows a gross level comparison of the relative risk potential among the various structures. For example, it is evident that structural type I2 is a higher perceived risk than type U but it is not clear whether type I2 is of higher or lower risk than type H1. In order to classify the structural types according to fire risk potential, a single valued measure of risk is needed.

Table 7. Scaled Risk Measures

BOCA	Scaled Values		Risk Value	Risk Category
	Consequence	Probability		
I2	0.43	1.00	0.72	Maximum Risk
A4	1.00	0.14	0.57	Maximum Risk
A2	1.00	0.07	0.53	Maximum Risk
A1	1.00	0.00	0.50	Maximum Risk
A3	1.00	0.00	0.50	Maximum Risk
R1	1.00	0.00	0.50	Maximum Risk
M	0.43	0.22	0.33	High Risk
B	0.34	0.32	0.33	High Risk
H1	0.61	0.00	0.31	High Risk
H2	0.61	0.00	0.31	High Risk
H3	0.61	0.00	0.31	High Risk
H4	0.61	0.00	0.31	High Risk
R3	0.21	0.26	0.24	Moderate Risk
F1	0.43	0.00	0.22	Moderate Risk
S1	0.43	0.00	0.22	Moderate Risk
I3	0.43	0.00	0.22	Moderate Risk
R2	0.34	0.02	0.18	Moderate Risk
E	0.34	0.00	0.17	Moderate Risk
F2	0.34	0.00	0.17	Moderate Risk
I1	0.34	0.00	0.17	Moderate Risk
S2	0.34	0.00	0.17	Moderate Risk
R4	0.21	0.00	0.11	Low Risk
U	0.13	0.00	0.06	Low Risk

A simple additive weighting (SAW) function is used to convert the (Consequence, Probability) risk doublets to a single value of fire risk potential for each BOCA code. The form of this linear function is:

$$\text{Risk}_i = w_c C_i + w_P P_I \qquad (3)$$

where,

$\text{Risk}_i \equiv$ measure of risk, for $i = A1, A2, \ldots, M$;

$w_j \equiv$ weighting factor for $j = C, P, wC + wP = 1.0$;

$C_i \equiv$ measure of consequence, for $i = A1, A2, \ldots, M$; and,

$P_i \equiv$ measure of probability, for $i = A1, A2, \ldots, M$.

In this particular instance, the variables wC and wP are both set to 0.5 to reflect the equal importance given by the fire district to consequence and probability. Other communities may modify these weighting factors to accommodate different

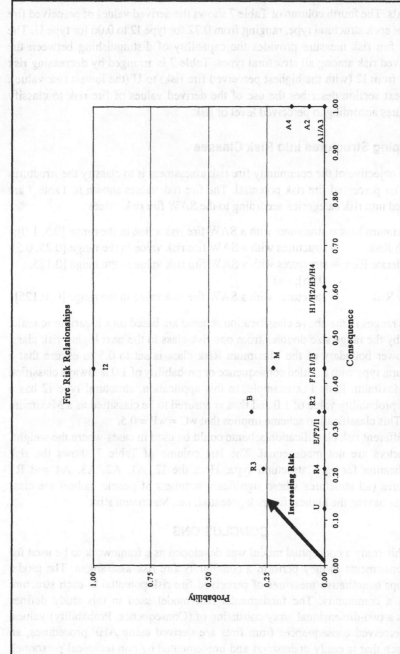

Figure 4. Fire risk potential.

contexts. The fourth column of Table 7 shows the derived values of perceived fire risk for each structural type, ranging from 0.72 for type I2 to 0.06 for type U. The single fire risk measure provides the capability of distinguishing between the perceived risk among all structural types. Table 7 is arranged by decreasing risk value, from I2 (with the highest perceived fire risk) to U (the lowest risk value). The next section describes the use of the derived values of fire risk to classify structures according to perceived level of risk.

Grouping Structures into Risk Classes

The objective of the community fire risk assessment is to classify the structures based on perceived fire risk potential. The fire risk values shown in Table 7 are grouped into risk categories according to the SAW fire risk values:

Maximum Risk ≡ structures with a SAW fire risk value in the range [0.5, 1.0];
High Risk ≡ structures with a SAW fire risk value in the range [0.25, 0.5];
Moderate Risk ≡ structures with a SAW fire risk value in the range [0.125, 0.25]; and
Low Risk ≡ structures with a SAW fire risk value in the range [0, 0.125].

The ranges in the above classification scheme are based on a logarithmic scale, whereby the range size doubles from one risk class to the next higher risk class. The lower boundary of the Maximum Risk class is set to 0.5 to ensure that a structural type with a scaled consequence or probability of 1.0 is always classified as a Maximum Risk. For example, in this application, structural type I2 has a scaled probability value of 1.0 and thus is ensured to be classified as a Maximum Risk. This classification scheme implies that $wC = wP = 0.5$.

A different risk classification scheme could be used in cases where the weighting factors are not made equal. The last column of Table 7 shows the risk classification for each structural type. Here the I2, A1, A2, A3, A4 and R1 structures (all structures where significant numbers of people gather) are classified as having the highest fire risk potential, i.e., Maximum Risk.

CONCLUSIONS

In this study a conceptual model was developed as a framework to be used for fire departments as they perform a community fire risk assessment. The model develops quantitative measures of perceived fire risk potential to each structure within a community. The fundamental risk model used in this study defines risk as a two-dimensional array consisting of (Consequence, Probability) values. The perceived consequences from fires are derived using AHP procedures, an approach that is easily understood and implemented by non-technical personnel. The probability of fire is found from statistical analysis of historical data. Thus, the derived values of fire risk reflect empirical data (i.e., the frequency of fire calls) and the context and expert opinion of the people most directly linked to a particular fire district (i.e., the fire officers and other community leaders). Although the specific model parameters and results reflect the specific perceptions and conditions of a particular community, the models developed are easily changed to accommodate different communities and contexts.

The risk model was used for a fire district in the City of Hampton, Virginia. The results achieved during the implementation phase showed that the model is easily adaptable, and provides a mechanism for handling the inherent subjectivity involved in the fire risk assessment. The work performed for the City of Hampton is a starting point for further study in this area. Every structure in the fire district may be assigned a risk classification according to its BOCA code (e.g., all B structures assigned as High Risk). The risk classification of specific structures known to have singularly distinctive characteristics may then be adjusted accordingly. These risk classifications may subsequently be input to a Geographical Interface System (GIS) to develop risk demand zones within the city. Each risk classification can be color coded (e.g., all Maximum Risk structures shown in red) to highlight their location within the community. Areas of higher relative risk will thus be easily identified and their relative concentration noted. These demand zones may serve as useful input for establishing policy and making resource allocation decisions. In addition, these demand zones are under consideration as a potential input to a software simulation model of the fire district.

REFERENCES

1. *Fire and Emergency Service Self-Assessment Manual,* National Fire Service Accreditation Task Force, International Association of Fire Chiefs, 1997.
2. *Risk Management: Concepts and Guidance,* Defense Systems Management College, Ft. Belvoir, Virginia, 1990.
3. W. Phillips, Simulation Models for Fire Risk Assessment, *Fire Safety Journal, 23*:2, pp. 159-169, 1994.
4. R. Bruegman, R. Coleman, and P. Brooks, Self-Assessment: Safeguarding the Future, *Fire Engineering, 150*:3, pp. 83-94, June 1997.
5. R. Coleman, Risk Hazard and Value Evaluation: A Perspective, *Fire Engineering, 150*:2, pp. 93-100, March 1997.
6. *Standards of Response Coverage for California Fire Departments,* Association of California Fire Departments, Sacramento, California, 1997.
7. *National Building Code,* Building Officials Code Administrators, Country Club Hills, Illinois, 1996.
8. P. Korhonen, H. Moskowitz, and J. Wallenius, Multiple Criteria Decision Support—A Review, *European Journal of Operational Research, 63,* pp. 361-375, 1992.
9. T. Saaty, How to Make a Decision: The Analytic Hierarchy Process, *Interfaces, 24*:6, pp. 9-43, 1994.
10. A. Fernandez, Expert Choice—A Software Review, *OR/MS Today, 23*:4, pp. 80-84, August 1996.
11. *Expert Choice,* Expert Choice, Inc., Pittsburgh, Pennsylvania, 1996.

CHAPTER 7

Calculation of Axial Forces Generated in Restrained Pin Ended Steel Columns Subjected to High Temperatures

F. A. Ali and D. J. O'Connor

Current method of calculating axial forces generated in axially restrained pin ended steel columns subjected to high temperatures considers two types of axial deformations, column thermal expansion, and column extracting under loads due to deterioration of steel properties (Young modulus). This chapter represents a modification to the current method by considering a third type of displacements. As the axially loaded pin ended column starts to lose its stability and deforms laterally, a shortening in the vertical height of the column takes place. The chapter investigates this problem and represents a modification to the current method. In a comparison with experimental data the modified method showed a better agreement and produced more realistic curves particularly in stages close to column failure. The chapter includes also a comparison between theoretical values of axial forces calculated by the current and the modified method.

The performance of steel columns under high temperatures has a significant effect on the fire safety of steel buildings. The exposure to elevated temperatures deteriorates the mechanical properties of the steel including its Young modulus and plastic deformations may develop. Among the factors that affect the column performance during fire is the restraint condition. When a column is heated a thermal expansion takes place. If this thermal expansion is restrained axial forces will generate. These forces increase the loading level and accelerate the column failure. In buildings, this restraint is imposed by the adjacent parts of the structure. The stiffness of the adjacent elements and their temperature difference relative to the column has a significant effect on the value of the induced axial forces. To analyze and predict the performance of steel columns in fire situations, various studies were performed in the past. Anderberg et al. [1], Bennetts et al. [2], Burgess et al. [3], Franssen et al. [4, 5], Olawale et al. [6], Bailey et al. [7] have investigated this problem considering various aspects [8]. Among the performed studies, references [2, 9, 10] have focused on determining the axial forces generated in restrained steel columns. Studying the available references shows that a significant factor, which can affect the calculations, has not been considered previously. When a column is loaded and heated it starts to lose its elastic stability after a period of time. During the instability phase the lateral displacement of the column develops rapidly. This displacement induces axial shortening in the vertical length between the top and the bottom of the pin ended column as shown in Figure 1. This vertical shortening increases with lateral displacement development until failure by buckling takes place and the vertical shortening reaches its maximum value (Figure 1). The current method does not take into account this factor which can affect the calculated values of the generated axial forces particularly in stages close to failure.

The objective of this article is to present a modified method for determining the value of axial forces generated in axially restrained pin ended steel columns subjected to elevated temperatures by taking into account the column vertical shortening due to buckling. This would provide more accurate fire safety calculations and more precise prediction of failure temperatures.

CURRENT METHOD

Under high temperatures a column thermal expansion can take place. If the axial expansion is restrained axial force P_r will be generated in the column. The value of the force P_r is dependent on the stiffness of the surrounding building parts that contribute to restraining the column expansion during fire. If the stiffness of those elements is replaced with a spring of stiffness K_s (Figure 1) then the value of P_r can be calculated using basics of structural mechanics:

$$P_r = \Delta_{TOTAL} \cdot K_s \tag{1}$$

where Δ_{TOTAL} - is the total change in the column length during heating:

$$\Delta_{TOTAL} = \Delta_{THERM} - \Delta_{DET} - \Delta_{REST} \tag{2}$$

The first term on the RHS of equation (2) represents the free thermal expansion of the column:

$$\Delta_{THERM} = \alpha \Delta T L \tag{3}$$

where ΔT = temperature difference, L = column length, and α = coefficient of thermal expansion. The second term in equation (2) is the shortening in the column length under the effect of the applied loads attenuated by material property deterioration at elevated temperatures:

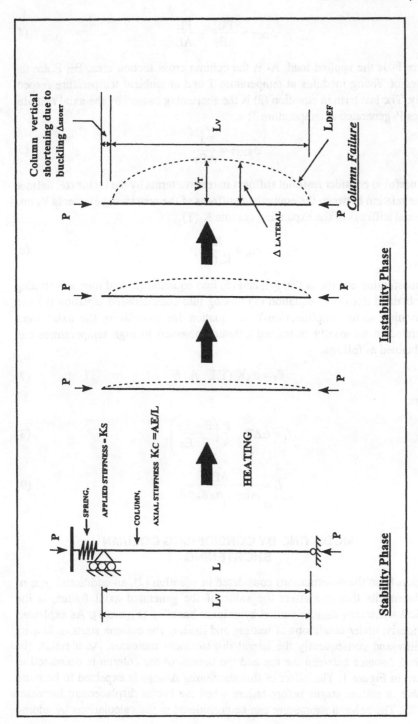

Figure 1. Definition of the considered problem.

$$\Delta_{DET} = \frac{PL}{AE_T} - \frac{PL}{AE} \tag{4}$$

where P- is the applied load, A- is the column cross section area, E_T, E are the values of Young modulus at temperature T and at ambient temperature respectively. The last term in equation (2) is the shortening caused by the axial restraint forces P_r generated at temperature T:

$$\Delta_{REST} = \frac{P_r L}{AE_T} \tag{5}$$

It is useful to consider restraint stiffness in relative terms by the factor α_K, defined as the relation between the equivalent stiffness of the restraining elements K_s and the axial stiffness of the expanding column $K_c(T)$:

$$\alpha_K = \frac{K_s}{K_c(T)}, \tag{6}$$

By substituting equations (3), (4), and (5) into equation (2) and then substituting the obtained Δ_{TOTAL} in equation (1) (taking into consideration equation (6) and performing some simplifications) an equation for calculating the axial force generated in an axially restrained column subjected to high temperatures can be obtained as follows:

$$P_r = \alpha_K K_c(T)L \cdot f_1 \cdot f_2 \tag{7}$$

where:

$$f_1 = \alpha \Delta T - \frac{P}{A}\left(\frac{E - E_T}{E \cdot E_T}\right), \tag{8}$$

$$f_2 = \frac{AE_T}{AE_T + \alpha_K K_{c(T)}L} \tag{9}$$

MODIFYING BY CONSIDERING COLUMN SHORTENING

Apart from the deformations considered in equation (2), an additional type of displacements that can affect the value of the generated axial forces, is the vertical shortening Δ_{SHORT} caused by column buckling (Figure 1). As explained previously, under conditions of heating and loading the column starts to lose its stability and consequently the lateral displacement increases. As a result, the vertical distance between the top and the bottom of the column is decreased as shown in Figure 1. The effect of this shortening Δ_{SHORT} is expected to be more notable in mature stages before failure when the lateral displacement increases rapidly. The column shortening can be considered in the calculations by adding

Δ_{SHORT} to equation (2) and following the same deriving procedure. The conclusive equations are similar to equations (7), (8), and (9) with a slight difference:

$$P_r = \alpha_K K_c(T)L \cdot f_1 \cdot f_2 \,; \qquad P_r \geq 0 \tag{10}$$

where:

$$f_1 = \alpha \Delta T - \frac{P}{A}\left(\frac{E - E_T}{E \cdot E_T}\right) - \Delta_{SHORT}, \tag{11}$$

$$f_2 = \frac{AE_T}{AE_T + \alpha_K K_{c(T)}L} \tag{12}$$

To calculate the value of the shortening Δ_{SHORT} introduced in equation (11) let the vertical distance between the top and the bottom of the column be L_v, the length of the column be L and the lateral displacement be $\Delta_{LATERAL}$ (Figure 1). The value of the axial vertical shortening Δ_{SHORT} can therefore be found using the relation:

$$\Delta_{SHORT} = L - L_v \tag{13}$$

From Figure 1 it is obvious that L_v decreases as the column starts to lose its lateral stability. Therefore, the following boundary conditions can be applied:

$$\text{if} \qquad \Delta_{LATERAL} = 0; \qquad L_v = L, \tag{14}$$

$$\text{and when} \qquad \Delta_{LATERAL} > 0; \qquad L_v < L \tag{15}$$

The other important boundary condition is that the length of the column remains constant before and after buckling (traditional strain due to loading and heating is considered previously):

$$L = L_{DEF} \tag{16}$$

where L_{DEF} = the length of the bent column. In order to calculate the value of L_v in equation (13), a single half-sine wave column deforming shape is assumed [11]:

$$\Delta_{LATERAL} = v_T \text{Sin} \frac{\pi x}{L_v} \tag{17}$$

where v_T = the lateral displacement at the mid height of the column. The length of the arc L_{DEF} (Figure 1) which represents the deformed column can be found utilizing equation (17):

$$L_{DEF} = \int_0^{L_v} \sqrt{\left(1 + \frac{v_T^2 \pi^2}{L_v^2} \text{Cos}^2\left(\frac{\pi x}{L_v}\right)\right)} \, dx \tag{18}$$

It is obvious that the length of the deformed column L_{DEF}, if calculated using equation (18), will be larger than the length of the straight column L. But this contradicts the boundary condition in (16). Therefore, for a given lateral displacement v_T, determining the vertical length L_v requires an iterative process which consists of two steps.

Step 1

Substitute v_T in equation (18) and find L_{DEF} assuming that $L_v = L$.

Step 2

Readjust (reduce) L_v value in equation (18) until, implementing the boundary condition in (16), when L_{DEF} (calculated in step 1) on the LHS of equation (18) becomes equal to L.

The procedure requires numerical solution of equation (18) as the integral can not be solved analytically. The Simpson's rule can be used with a reasonable accuracy. The value of the length L_v calculated from step 2 can then be substituted in equation (13) to find Δ_{SHORT}. By determining Δ_{SHORT} the value of the generated axial force can be found using equations (10), (11), and (12).

The procedure should be repeated for each temperature increment. For a column failed at 430°C a division of 100°C can be adopted. However, smaller increments should be used between 400°C and failure temperature.

COMPARISON WITH EXPERIMENTAL DATA

To assess the modified method, a comparison with test data [9, 10] was carried out. The test data used in the comparison is a part of an undergoing joint research with the University of Sheffield, UK [9]. The considered specimen is a pin ended column 127x76UB13 [12], 1800 mm length subjected to axial restraint value 37 kN/mm. This is equivalent to a restraint degree $\alpha_K = 0.2$. The column was heated until failure took place by buckling about the weak axis. Table 1 represents the experimental values of the generated axial force and the theoretical values calculated using the current method (equation (7)) and the modified method (equation (10) and the discussed steps). Table 1 also illustrates the development of the lateral displacement and vertical shortening Δ_{SHORT} calculated by the modified method. Curves representing the axial force values are shown in Figure 2. Table 1 emphasizes the moments just before failure (T > 500°C) where the column shortening effect is more crucial.

The second comparison was performed with specimen TP112 [10] of a joist section IPE80x46 subjected to restraint degree $\alpha_K = 0.042$ and loading level 0.61 of its capacity according to BS5950 [12]. The column had a 1600 mm length and a slenderness $\lambda = 136$. Figure 3 shows the experimental curve of the axial forces generated in the column and the calculated curves using the current and the

Table 1. Experimental and Theoretical Values of Axial Force
Generated in Column UB13

Temperature °C	Test Results (kN)	Current Method (kN)	Modified Method		
			Axial Force (kN)	Column Lateral Displacement (mm)	Column Shortening ΔSHORT (mm)
20	0	0	0	0.200[a]	0
100	15.81	53.7	53.58	0.221	0
200	79.2	118.91	118.41	0.263	0
300	146.2	181.05	180.21	0.340	0
400	226.74	239.19	237.56	0.586	0.001
500	268.39	291.81	290.08	1.176	0.001
510	264.4	294.31	292.32	3.320	0.012
515	263.4	295.90	291.92	6.240	0.045
520	257.8	296.37	50.69	69.67	6.59
521	256.7	296.56	0	200.8	50.8
528[b]	54.43	297.95	0	200.8	50.8

[a]Initial curvature
[b]Failure; 0.1% iteration accuracy was used in the numerical solution.

modified method. It is clear from Figure 3 that the modified method gives a reasonable agreement with the test particularly upon failure where a 235°C failure temperature was predicted and the experimental failure temperature was 233°C.

COMPARISON BETWEEN THE TWO METHODS

A comparison between theoretical values calculated by the current and the modified methods was performed for various restraint degrees. A pin ended steel column, section 127x76UB13, 1800 mm length is considered in the comparison. It has been assumed that the column is loaded with 49 kN representing 0.2 of its capacity according to BS5950 [12], and that the column has an initial curvature 0.25 mm. Three degrees of restraint were imposed against the thermal expansion of the column 0.1, 0.2, and 0.3. Figure 4 shows the curves representing the calculated values of the generated axial forces in the column. Examining Figure 4 shows that the modified method has produced more realistic curves and predicted a specific failure temperature. Curves produced by the modified method represent the traditional shape of axial force development obtained from previous experiments [9, 10] (Figures 2 and 3).

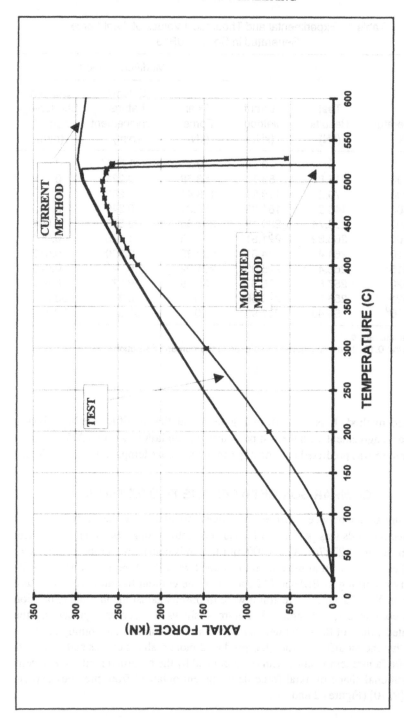

Figure 2. Experimental values of axial forces generated in the restrained column UB13 and values calculated using current and modified methods.

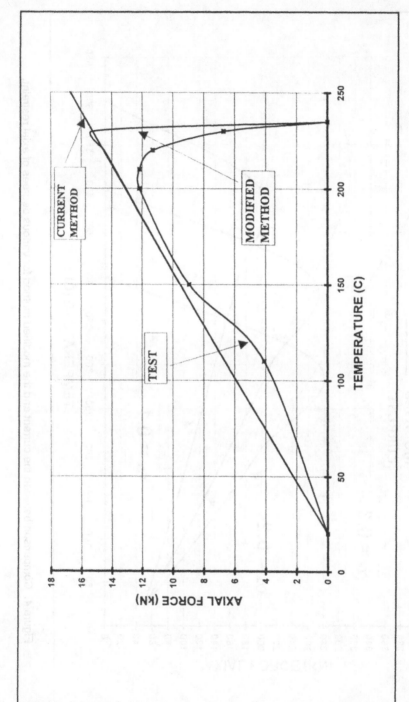

Figure 3. Experimental values of axial forces generated in the restrained column TP112 and values calculated using current and modified methods.

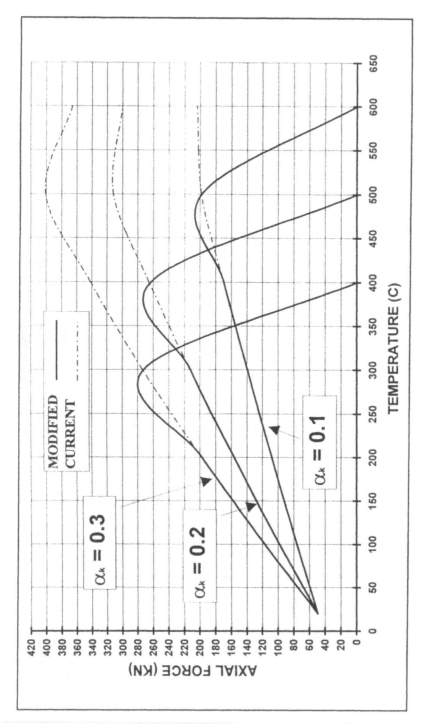

Figure 4. Comparison between the current and the modified method for various degrees of axial restraint.

DISCUSSION

As mentioned previously, the effect of the vertical shortening of a pin ended column is more distinct in stages just before failure. This is when the column loses its stability and the lateral displacement rapidly increases. This can be noticed from the data presented in Table 1. The values of the axial forces calculated by the modified method are very close to those calculated by the current method in early heating stages (20°C <T< 400°C). But the difference between the two methods becomes more evident in temperatures higher than 400°C as the lateral displacement increases quickly and consequently the vertical shortening of the column increases. Perhaps the most significant capability of the modified method is that it enables to predict a particular failure temperature while the current method cannot do so. For the two cases considered, the predicted failure temperatures were 521°C and 235°C while the test values were 528°C and 233°C [9, 10] respectively. By examining Figures 2, 3, and 4 it can be noticed that the modified method has produced more realistic curves than the current. Despite that the modified method is applicable to pin ended columns it is still appropriate to other end conditions where further investigation is necessary.

CONCLUSIONS

1. A modified method to calculate the values of axial forces generated in axially restrained pin ended steel columns subjected to high temperatures is presented. The method predicts more realistic values of axial forces particularly in mature stages before failure.
2. The comparison shows that the modified method gives satisfactory agreement with test data.
3. The method is capable to predict a particular failure temperature.
4. The presented method is more practical for the elastic instability phase and enables determining the column shortening due to lateral displacement (buckling).

REFERENCES

1. Y. Anderberg, N. E. Forsen, and B. Aasen, *Measured and Predicted Behaviour of Steel Beams and Columns in Fire*, Fire Safety Science—Proceedings of the First International Symposium, USA, pp. 259-269, 1985.
2. I. D. Bennetts, C. C. Goh, A. J. O'Meagher, and I. R. Thomas, *Restraint of Compression Members in Fire*, Report MRL/PS65/89/002, September 1989.
3. I. W. Burgess and S. R. Najjar, A Simple Approach to the Behavior of Steel Columns in Fire, *Journal of Constructional Steel Research, 31*:1, pp. 115-134, 1994.
4. J.-M. Franssen and J. C. Dotreppe, Fire Resistance of Columns in Steel Frames, *Fire Safety Journal, 19*:2-3, pp. 159-175, 1992.

5. J.-M. Franssen, J.-B. Schleich, and L.-G. Gajot, Simple Model for the Fire Resistance of Axially Loaded Members According to Eurocode 3, *Journal of Constructional Steel Research, 35*:1, pp. 49-69, 1995.

6. A. O. Olawale and R. J. Plank, Collapse Analysis of Steel Columns in Fire Using a Finite Stripe Method, *International Journal for Numerical Methods in Engineering, 26*:12, pp. 2755-2764, 1988.

7. C. G. Bailey, M. Wadee, K. Baltzer, and G. M. Newman, *The Behavior of Steel Columns in Fire*, Report RT524, Version 1, submitted by The Steel Construction Institute to the Department of Environment, United Kingdom, March 1996.

8. P. J. E. Sullivan, M. J. Terro, and W. A. Morris, Critical Review of Fire Dedicated Thermal and Structural Computer Programs, *Journal of Applied Fire Science, 3*:2, pp.113-135, 1993-94.

9. F. A. Ali, I. W. Simms, and D. J. O'Connor, Behaviour of Axially Restrained Steel Columns during Fire, *Proceedings of the Fifth International Fire and Safety Conference*, Melbourne, Australia, March 1997.

10. W. I. Simms, D. J. O'Connor, F. Ali, and M. Randall, An Experimental Investigation on the Structural Performance of Steel Columns Subjected to Elevated Temperatures, *Journal of Applied Fire Science, 5*:4, December 1996.

11. W. T. Marshall and H. Nelson, *Structures*, Longman Group, London, pp. 420-425, 1990.

12. British Standard Institution BS5950, *The Structural Use of Steel Work in Building, Part 1: Code of Practice for Design in Simple and Continuous Construction*, BSI, London, 1990.

CHAPTER 8

A Case Study of Fire Protection in Large Stadia: The San Antonio Alamodome

Glenn P. Corbett

Large stadia present unique fire protection challenges. In order to ensure the life safety of thousands of patrons, the building design must incorporate and integrate a variety of fire protection features including fire suppression, smoke management, fire detection and alarms, and a means of public address. The Alamodome, recently constructed in San Antonio, provides a case study of the life safety and fire protection in structures such as these. Current codes and standards typically fall short of addressing the fire and life safety concerns presented. "Design fires" were created to study how the building and its occupants will respond to a particular fire scenario. Life safety issues other than fires were also considered and evaluated. Unique fire protection and life safety systems and features are incorporated into the design and subsequently tested. The preparation and ongoing utilization of a Life Safety Evaluation for the life of the building ensures that life safety issues will continue to be addressed.

In 1989, the citizens of Bexar County, Texas, voted to increase the local sales tax by 1/2 cent to raise the necessary funds for construction of a large, 182 million dollar multi-purpose dome facility in downtown San Antonio. The "Alamodome" (as it was later named) was built to provide a single "large-area" convention space (including trade shows) not available in the city's existing convention center, a facility capable of handling large entertainment events, as well as a building configured to offer the opportunity of hosting professional sporting events, such as *San Antonio Spurs* basketball and possibly football, hockey, among others.

This chapter will focus on the fire protection aspects of the facility and the process used to review and approve the building. Since the author was the

Administrator of Engineering Services for the San Antonio Fire Department at the time of design and construction of the facility, the article highlights those issues of importance related to code-compliance.

ALAMODOME

In order to provide an understanding of the layout and use of the facility, consider the following Alamodome facts:

- It measures approximately 450 feet long by 250 feet wide (18 acres).
- It is placed on a site of fifty-seven acres with the remainder of the site used for parking, transit facilities, etc. The site is very "tight," with a freight railroad and Interstate 37 directly next to the Alamodome on either side.
- It has four interior levels with the field level (playing field) at the base, followed by the plaza (entry) level, club level, and upper concourse level.
- The vertical distance from playing field to the roof deck is 172 feet.
- It has seating capacity for 65,000 people but can be expanded to 73,200 (Superbowl requirements).
- It is the home of the *San Antonio Spurs* basketball team and is capable of handling most other sporting events with the exception of baseball. It has one of the few "double" side-by-side full-scale hockey rinks in the country.
- It provides 160,000 square feet of exhibit space and 30,000 square feet of meeting space for conventions.
- It is capable of handling a variety of non-sporting events such as "monster truck" events and "tractor pulls," religious ceremonies, contemporary and rock music concerts (there are a multitude of seating configurations including a "stage in the round").
- It has a unique roof structure including two massive steel (200 ton, 376 foot long) bowstring trusses. Four, 300 foot tall concrete "masts" at each corner of the building carry the roof load through the use of cables. Even though the building uses the name Alamo*dome,* in reality, the structure is not a "dome."

CODE ISSUES

As soon as the sales tax referendum was passed, the San Antonio Fire Department began an informal survey of existing "dome" and arena facilities. Several fire departments across the country were contacted to learn of their experiences with the domes in their jurisdictions. Experts in the field of fire protection and life safety in these types of facilities were called upon. (A literature search was also conducted, but very few articles were uncovered.) The following conclusions were drawn from these contacts:

- No single fire protection code or other nationally recognized standard existed that cover *all* of the issues of concern in these types of buildings.
- Issues of "general" life safety (people movement, etc.) were as important as fire protection issues.
- The importance of fixed fire protection equipment became even more important when viewed in context of the limited site access for San Antonio firefighters—it was estimated that it may take upwards of twenty minutes for firefighters from dispatch to arrive at the actual location of the fire after "fighting" their way into the building through the throngs of patrons attempting to leave.
- Certain fixed fire protection equipment proved unreliable and had to be modified.
- A fire department representative needed to be assigned full-time, and a "dome" employee assigned responsibility for fire protection issues in the dome *for the life of the building* to ensure for smooth operations.
- Facilities of this type require significant resources in terms of ongoing inspections and fire marshal presence at various dome "events."

San Antonio's construction and fire-safety codes in effect at the time of the Alamodome's design were the *1985 Uniform Building Code* published by the International Conference of Building Officials (ICBO) (enforced by the San Antonio Building Inspections Department) and the *1985 Uniform Fire Code* published by the Western Fire Chiefs/ICBO (enforced by the San Antonio Fire Department). Local amendments were also in effect as well.

A review of San Antonio's adopted codes provided little guidance for a building of this type. It was readily evident that additional design standards and experts would need to be consulted in order to assure the safety of occupants as well as firefighters.

The following general concerns began to emerge for San Antonio's plan reviewers and inspectors:

- Access by firefighters would be greatly hindered by the constraints of the site. The final design only allowed for fire apparatus to access the north and south of the structure. These two access areas were *on two different levels* (plaza level on the north, field level on the south).
- Automatic fire sprinklers appeared to be of questionable value over the center of the dome playing field/exhibit floor, due to delayed activation time (sprinklers at the roof level would be nearly 170 feet above the floor allowing a large body of fire to develop) and causing water droplet plume penetration problems.
- Smoke management would be a key issue.

- Egress capacity and travel distance would not meet the requirements of the *1985 Uniform Building Code* (hereafter UBC) by a large margin.
- Crowd control and people movement were significant life safety concerns.

It was evident that additional codes/standards would be necessary to address the concerns enumerated above. In particular, two documents outside of San Antonio's adopted regulations had a tremendous impact on the Alamodome design. They were NFPA 101, *The Life Safety Code, 1988 Edition* and the proposed NFPA 92B, *Guide for Smoke Management Systems in Malls, Atria, and Large Areas*. NFPA 92B was a brand new document still in development during the initial stages of design and was not actually promulgated by the NFPA until the beginning of 1991. Since the adopted fire and building codes permitted the use of other "nationally recognized standards of practice," NFPA 101 and the proposed NFPA 92B were utilized to review and inspect the facility.

In addition, due to the complexity and unique nature of this structure, additional technical expertise would be needed on the plan review/inspection team (i.e., the San Antonio Fire and Building Inspections Departments) in the areas of life safety. As the review/approval authorities, the City of San Antonio wanted to have a life safety consultant identify problems and propose solutions directly to City Fire and Building Officials. Additionally, experts on smoke management were retained by the design team and mechanical contractor to (among other duties) address issues raised by the city authorities.

GENERAL FIRE PROTECTION DESIGN CONSIDERATIONS

Overall, the building was classified as "Type I—F.R." construction under the Uniform Building Code (UBC). In addition, since the UBC itself had been amended by San Antonio to require *all* structures to comply with the "high-rise" provisions, a "high-rise" package was required for the Alamodome. These requirements include:

- automatic sprinkler protection throughout
- firefighters communication system
- public address system
- waterflow alarms
- automatic detection in mechanical rooms and similar areas
- smoke management features
- fire control room

Additional fire protection requirements in the UFC and UBC include:

- adequate fire hydrants and fire flow
- cooking hood fire protection
- fire extinguishers throughout
- a Class III standpipe including 1-1/2″ hose stations in "stage" areas
- elevator lobby smoke detection
- appropriate structural fire protection
- two-hour-rated stair and elevator shafts, one-hour-rated corridors
- elevator recall features

Although many of the fire protection features could readily be included in the Alamodome, some were not advisable in this building. It was essential that all code provisions be reviewed and studied to determine whether or not they made sense in this facility. The complexity of the building and the desire to make the fire protection "state of the art" made the review of the facility difficult. An integrated approach using the codes, sound fire protection engineering principles, and research would be necessary.

LEVELS OF PERFORMANCE AND DESIGN FIRES

As the building design began to take shape, three specific major concerns began to emerge—the provisions for smoke management, the structural integrity of the unprotected roof structure, as well as life safety/egress. Additionally, the critical interrelationship between these three concerns made them a complex problem to solve.

It also became apparent that "outside" fire protection experts would be needed for certain aspects of the project. Three independent fire protection "smoke" consultants were retained by the design team and mechanical contractor to evaluate the proposed smoke management design and the related roof integrity issue as well as to answer questions raised by the San Antonio Fire and Building Inspections Departments. A life safety consultant was retained by the city itself as previously cited.

Before the evolving building design could be evaluated, it was necessary to first establish a level of performance and a set of design fires against which the proposed design would be reviewed. NFPA 101 established one of these performance goals through its requirement for a "smoke-protected" seating area when more "liberal" egress capacities (narrower aisle widths, etc.) are utilized [1]. Effectively, this requirement establishes that the "smoke layer" must be at least 6 feet above all walking surfaces. It was the intent to provide this level of protection until all occupants had exited the structure.

Inherent to this egress requirement is the obvious desire to contain a fire to a size incapable of violating the smoke layer criteria just described. The study of the smoke layer would also be critical to establishing the ceiling (roof) layer temperature and verifying that the temperatures developed would not be destructive to the roof's structural members.

Generally, the smoke consultants agreed that two fire scenarios should be studied in detail—a large fire on field level as well as a fire involving a concession booth on the concourse walkway behind the seating areas. The consultants parted ways, however, in estimating the fire size itself and the method of calculating the rate of smoke production rate and resultant smoke exhaust capacity needs.

The field itself is used for a variety of events (trade shows, concerts, tractor pulls, etc.). Two smoke consultants felt that a heat release rate of 5 MW was appropriate for a typical trade show exhibit booth fire or other field level fire. The other consultant felt that a larger design was called for—a 20 MW fire. It was felt by this consultant that shielded fires such as under a concert stage/platform set-up could develop at field level. The first two consultants also found that 5.3 MW was also appropriate for a concession fire while the other consultant felt a 5 MW fire was more appropriate.

The Alamodome's mechanical design team had provided the building with a substantial smoke exhaust system. The seating area itself was provided with thirty-two fans with a total exhaust capacity of 1,450,240 cfm. The concourses were provided with an exhaust capacity of 1,424,000 cfm utilizing twenty fans.

Calculations were performed for both a field level fire and concession fire. Specifically, all of the smoke consultants wished to answer two questions regarding the field level fire: 1) were the roof structural members in danger of being damaged due to the "ceiling layer" temperature, and 2) would the rate of smoke production exceed the design capacity of the smoke management system? The same question concerning the capability of the smoke management system in the concession area also had to be answered.

If we consider the worst-case design fire on the field level and assuming the following:

- the ambient temperature at $t = 0$ is 70°F
- space temperature variation with height at $t = 0$ is 0.06°F/ft
- a *conservative* maximum safe temperature at roof of 115°F
- bottom of the smoke layer not to descend below 120 feet above field level (the top of upper concourse seating area)
- convective heat output is (Q_c) is 0.7Q
- the fire size is 20 MW (Q)

The apparent flame height is given as

$$z_1 = 0.533Q_c^{2/5}$$

with $Q = 20$ MW, $z_1 = 24$ feet.

The rate of smoke production (m_p) is given as:

$$m_p = 0.022 \, Q_c^{1/3} \, z^{5/3} + 0.0042 Q_c$$

for t = 0 and z = 180 feet, m_p = 3,101 lbs./sec;
for z = 120 feet, m_p = 1,607 lbs./sec.

Therefore, the exhaust requirement in terms of capacity is 2,658,000 cfm when z = 180 feet and 1,377,000 cfm when z = 120 feet.

With this analysis it is shown that the initial smoke management system design capacity is appropriate for a typical field level fire while maintaining the ceiling layer temperature at or below 115°F.

"WATER CANNONS" AND THE AUTOMATIC SPRINKLER SYSTEM

Critical among the fire protection systems is the structure's automatic sprinkler system. This system is essential in keeping fires to a manageable size so that other fire protection features (i.e., smoke management) will perform properly. Another significant concern was the response time of "standard" sprinklers to a field level fire. While automatic sprinkler protection was appropriate for most areas of the Alamodome, it made little sense to require "standard" sprinklers at the roof deck over 160′+ above the field level. Previous tests conducted several years ago established that standard sprinklers could be useful for exhibition halls with ceilings as high as 50 feet, but with a resultant large fire area developing at floor level. In the case of the Alamodome, the "ceiling" height was three times as tall as the tests that had to be conducted [2]. Since it was not desirable to develop such a large fire at field level and have to wait a considerable amount of time for standard sprinklers to activate (not to mention plume penetration problems, etc.), another means of activating a water-based suppression system applying water to a fire in this area would be necessary.

Research was conducted as to the type of protection provided for similar domed stadia. Most had no protection for the playing field/exhibit floor. One facility had a water cannon system utilizing high flow monitor nozzles. This system was originally automated, using projected beam detectors for actuation. These detectors, however, apparently proved unreliable because of problems maintaining proper beam alignment as large crowds of people moving in and out of the stadium would "shake" the seating structure (upon which the detectors were mounted) moving the beam off of its target.

It became apparent that some type of water cannon would be necessary for the Alamodome. The monitors selected were to each be capable of flowing 2,000 gpm at 125 psi, full 180 degree travel within 18 seconds, and capable of ±75 degree vertical aiming, and applying a minimum of .16 gpm/square foot water density. They were also to be capable of being actuated automatically or manually as well as being operated remotely in the security center (see Figure 1). Six

Figure 1. The water cannon control panel in the security center.

monitors were to be installed along the edge of the mezzanine level—three on each side of the field.

Additionally, it was desired to have an automated system due to several concerns, including the need for "after hours" protection of the building when a trade show is in place. The type of detection equipment to activate the monitors was limited—most types of detectors would have a difficult task responding quickly due to their location at the roof deck. In addition, the plume width at the roof would be very wide and not allow for a more precise determination as to the fire location at floor level.

Flame detectors were selected as the water monitor detection system. Ultraviolet/infrared detectors were utilized for the task. They were placed at roof level so that all areas of the field were "covered." They were also zoned to divide the floor into six sections. The flame detectors were designed to select the two closest water cannons and activate them after an appropriate "countdown" delay.

Automatic sprinkler protection was installed throughout the rest of the Alamodome, including in the "sloping" roof structure above the upper level seating area. The office areas, meeting rooms, concourses, loading dock, bathrooms, and other typical areas were protected with standard protection in accordance with NFPA 13: *The Standard for the Installation of Automatic Sprinklers*. Special hazards such as the concession storage area and the artificial turf storage racks, however, called for different standards: NFPA 231: *General Storage* and NFPA 231C: *Rack Storage of Materials*, respectively, were used for the sprinkler system design in these areas.

The standpipe system is another important piece of equipment in the facility. Standpipe hose valves and risers were required in each stairwell of the building. It was recognized that, during conventions, trade booths on the exhibit floor would be located several hundred feet away from the closest standpipe outlet—a special design was called for.

Since 1-1/2" handlines would be called on to extinguish a trade booth fire, hose "stations" would need to be located throughout the exhibit area (see Figure 2). And since there were no columns to attach the stations to, it was decided to locate them in the floor. Metal floor boxes with lids were spaced throughout the floor. Inside of each box is 100' of hose and nozzle on a reel as well as a portable fire extinguisher. It was anticipated that lightweight plastic pylons would be placed on top of the boxes during trade shows to identify their location and prevent them from becoming blocked.

LIFE SAFETY CONCERNS

The complex nature and large occupant load of the Alamodome posed significant life safety concerns for the San Antonio code officials. Outside assistance was necessary, specifically to address people movement, egress design, human behavior emergencies, and other related issues. A life safety consultant was hired by the City of San Antonio to assist its code officials in addressing these concerns.

Due to the utilization of "reduced" egress requirements (permitted in conjunction with smoke-protected seating), the NFPA 101: *Life Safety Code*—1988 also required the preparation of a Life Safety Evaluation (LSE) [1]. This evaluation included the preparation of a report detailing the life safety strengths and weaknesses of the proposed facility design. It also includes a subsequent evaluation of the completed building while in actual use.

The LSE design review document covered a variety of issues, including several "non-fire" concerns. Specifically, the LSE addressed:

- an accounting of the nature of the events in the facility (ticketing practices, event purpose, time duration of event, etc.)
- occupant characteristics (familiarity with facility and event, socioeconomic factors, capabilities based upon age and physical abilities, etc.)
- management of the facility (use of operations manual, training, and supervision of personnel, etc.)
- emergency preparedness (disaster manual covering fires and severe weather, communication systems, etc.)
- building systems (fire protection, HVAC, structural soundness, weather protection, etc.)

The LSE identified hundreds of particular "points to ponder." Some of the more important egress issues included the Alamodome's four large "corner" stairs as well as travel distance criteria. Each of these issues (as well as many others) had to be studied and resolved.

The Alamodome relies heavily on four large corner stairs that "move" the majority of patrons from the upper levels of the structure for both normal and emergency egress purposes. These stairs, which were site-cast, are very wide (in excess of 20 feet wide each). The LSE pointed out the critical nature of step riser/tread dimension continuity (especially in light of them being site-cast), the critical need for proper placement of compliant side and intermediate handrails, and the ability to "direct" people in the stairs themselves using a public address system.

Travel distance to an exit is always a concern, especially in buildings of this type. Some of the distances from the fixed-seat areas of the upper levels of the building through the vomitories to the corner stairs exceeded the travel distance requirements specified in NFPA 101 (hence the importance of the concourse smoke management system). Non-fixed seats at field level (for concerts, etc.) also had to be "rearranged" to minimize travel distance problems.

The LSE is a "living" document. It must be reviewed and updated to address new types of events and changing conditions. As mentioned, observations of life safety conditions *during actual building use* must be periodically conducted.

TESTING THE WATER CANNONS AND SMOKE MANAGEMENT SYSTEM

Once the plans submitted for a building permit had been approved, Building and Fire Inspectors were assigned to coordinate inspections of the building. In addition to the "normal" inspections (fire alarm functional tests, hydrostatic tests, pump tests, etc.), a few critical tests had to be performed—a flow test of the water cannons and a test of the smoke management system (see Figure 2).

Figure 2. A test of a water cannon to verify operation.

The water cannons were tested using the following criteria:

- the application of at least .16 gpm/square foot throughout the field level
- full sweep coverage within 18 seconds as well as complete nozzle movements
- interface tests with detection system and remote control

The smoke management system was tested using the two fire scenarios developed during the design phase of the project—a 20 MW field level fire and a 5 MW concession fire on one of the concourses. The two simulated fires were created using a "seeded" propane burner flame. The 5 MW concourse fire was scaled down to .1 MW to minimize the direct flame impingement damage to concourse finishes (while still maintaining plume buoyancy to test the effectiveness of the concourse smoke management system (see Figures 3, 4, 5, 6, and 7).

Figure 3. A close-up of the 20 MW test apparatus showing the propane burners and "seed" trough.

Figure 4. The 20 MW test fire at the field level.

Figure 5. The tanker vehicle used to supply the propane test equipment.

Figure 6. The concourse level test fire.

Figure 7. Smoke being exhausted from the Alamodome during
the concourse fire test.

CONCLUSION

Domed stadium facilities present very unique challenges to both design teams
and code officials. Current fire protection codes have been outpaced by the
complexity of structures like these, necessitating the use of a hybrid of codes,
standards, fire test data, research documents, and engineering judgment. These
buildings also require the expertise of a variety of fire and life safety specialists
to ensure all issues are addressed.

REFERENCES

1. National Fire Protection Association, NFPA 101, *The Life Safety Code*—1988, Section 9-2.3.2.
2. W. A. Webb, Effectiveness of Automatic Sprinklers in Exhibition Halls, *Fire Technology, 4*:2, p. 121, 1968.

CHAPTER 9

A Systemic Approach to Fire Safety Offshore

J. Santos Reyes, A. N. Beard,
and P. J. Clark

Fire is probably one of the greatest hazards that can be encountered on an offshore platform. This is clearly seen in disasters such as the *Piper Alpha* disaster in July 1988. Fire Safety management on offshore platforms has been the subject of increasing interest since the publication of the Cullen report into the *Piper Alpha* fire. In order to be able to achieve and maintain an acceptable level of fire risk it is desirable to consider the system as a "dynamic whole." Very often, risk due to fire is analyzed by isolating "parts" that may produce fire. However, fire risk in a given situation cannot exist in complete isolation, but would be a result of an interaction of a number of "parts." Given this, a systemic approach to fire safety for an offshore platform is being pursued. This chapter gives an account of the work.

Fire risk is inevitably present in any industrial process involving hydrocarbons, including offshore installations. An accident due to fire may lead to a total loss of an offshore installation, as well as human life. Fire disasters such as at Manchester Airport in August 1985, *Piper Alpha,* and the more recent fire incident in the Channel Tunnel have illustrated that existing fire safety management methods may well be inadequate for dealing with such events. In order to be able to maintain an acceptable level of fire risk, a new way of looking at fire safety becomes necessary. Fire risk in a given situation cannot exist in complete isolation, but is a result of the interaction of a number of "parts." That is, fire safety must be considered as a product of the working of a "dynamic whole" [1, 2]. A systemic approach to fire safety is being pursued and it is described in this chapter.

121

MANAGING FIRE SAFETY OFFSHORE

Fire safety management on offshore platforms has been the subject of increasing interest since the publication of the Cullen inquiry in response to the *Piper Alpha* disaster [3]. This event demonstrated that the fire safety management was inadequate to deal with such a conflagration. It led to a shift from a prescriptive to a goal-setting approach. The prescriptive approach to safety in general is mainly based on procedures, rules, and guidelines intended for both regulators and offshore operators to produce acceptably safe operations. However, a goal-setting approach is based on safety goals or objectives associated with a risk assessment and a safety case for each operator's installation is developed; where a safety case is a document which demonstrates that an offshore installation is "tolerably safe," as far as the Health and Safety Executive is concerned.

Prescriptive Approach

A prescriptive approach to fire safety is based on guidelines, checklists, and rules. Weibye [4] lists some typical characteristics under this regime. They are:

- lack of top and line management commitment to safety;
- a group of staff responsible for the organization of safety;
- it is an adversarial approach, and
- generally hardware oriented.

Goal-Setting Approach

The goal-setting approach to safety has already been applied to the UK's onshore installations and to the Norwegian offshore facilities. The UK offshore installations were still under a prescriptive approach before the *Piper Alpha* fire. Following this, a public inquiry was conducted to investigate the facts and issues that led to this event. A main recommendation from this inquiry (Cullen report) was to replace the whole offshore safety regime (prescriptive approach) with a risk-based goal-setting approach and a "safety case" for each installation must be submitted to a regulatory body to be reviewed and approved. Under this approach, offshore operators are required to specify the standards to be used to satisfy the formulated safety goals. Three aspects should be contained or demonstrated in a safety case as recommended by Cullen [3]. They are:

1. The Safety Management System (SMS) of both the company and the offshore installation should be adequate to ensure that the design, operation, and equipment of the installation are safe.
2. Potential hazards and risk to personnel have been identified and appropriate controls provided.

3. In case of a major emergency, a Temporary Refuge (TR) for personnel on the installation should be provided, as well as safe evacuation, escape, and rescue procedures.

The company's safety management system is an essential ingredient of a safety case, since it addresses the design (conceptual and detailed) and the procedures (operational and emergency) of the operator's installations. Cullen recommended that:

> ... the Safety Management System (SMS) should set out the safety objectives, the system by which these objectives are to be achieved, the performance standards which are to be met, and the means by which adherence to these standards is to be monitored. It should draw on quality assurance principles similar to those stated in BS 5750 and ISO 9000 [3].

THE NEED FOR A SYSTEMIC APPROACH

A new way of looking at risk from fire is needed and this should include both "hard" and "soft" aspects. Hard aspects refer to information concerned with installations, equipment, personnel, and all those involved in the organization. Soft aspects refer to, inter alia, the opinions of staff and their perceptions of fire safety. In doing so, fire risk may be regarded as a product of a "system" [1, 2]. A system may be defined as a whole which is made of parts and relationships. Beyond that, some systems have a purpose. In general, it must be remarked that relationships among parts, pattern (systematically structured), and purpose are all acts of mental recognition rather than physical characteristics [5]. A system possesses a distinctive characteristic such as that emphasized by Ackoff [6].

> Performance of the whole is not the addition of the performance of the parts, but it is a consequence of the relationship between the performance of the parts. It is how performance relates, not how it occurs independently of the other parts. That is what systems thinking is about [6, p. 11].

The above quotation suggests that it is necessary to understand the relationships among the parts that may produce fire. Since, traditionally, fire risk is treated from an "isolation" point of view this will ultimately fail to fundamentally understand the nature of fire safety. That is, the cause of fire on an offshore installation may be found in the complexity of the relationships implicit in the design, procedures, equipment, environment, operations, etc.

In order to gain a full understanding and comprehensive awareness of fire risk in a given situation it is necessary to consider in a coherent way all the aspects that may produce fire [1]. In short, there is a need for a systemic approach to fire safety. "Systemic" means looking upon things as a system or as a dynamic whole; and "systemic" is not the same as "systematic." "Systematic" means being methodical or tidy. "Systemic" means seeing pattern and inter-relationship within a complex whole; to see events as products of the working of a system. A systemic approach should be systematic; but it is much more than this.

Approaches to Systems

As discussed above, fire safety is a complex issue that involves both "hard" and "soft" aspects. An approach needs to be adopted which is capable of dealing with this. An approach has been formulated which employs both the Viable System Model [7, 8] and the Failure Paradigm Method [9]. These approaches are described in the following sections. It is intended in this work to construct a Fire Safety Management System (FSMS) for offshore platforms. Essentially, such a system should be able to assess and maintain an acceptable level of fire risk.

The Viable System Model

The Viable System Model (VSM) was developed by Beer [7]. Viable systems are defined as those able to maintain a separate existence. This model is based on cybernetic principles where cybernetics is understood as pertaining to effective organization. Some cybernetic principles on which the VSM is based are:

- in a recursive organizational structure, any viable system is recursive; that is, it contains, and is contained in, a viable system;
- any viable system deals with its external environment. Hence, a viable system is influenced by its environment and this is influenced by a viable system. This relationship promotes learning;
- another principle on which the VSM is based is the Law of Requisite Variety. This means that only variety absorbs variety. Variety is a measure of complexity and expresses the number of possible states of a system.

According to Beer any viable system contains five necessary and sufficient systems that are strongly interconnected through a complex of information and control loops (Figure 1). They are the following:

System 1—System 1 parts are concerned with the implementation of the activities that produce an organization. They are themselves viable systems. They are autonomous and comply with VSM principles such as the recursion principle mentioned above. This can be seen in Figure 1 where a complete VSM fills the whole page. Two Systems 1, 1a and 1b (there could be more), are shown, each of them containing a complete viable system displayed at a 45 degree angle at a lower level of recursion. The square box at the top right-hand side of Figure 1 is the management of System 1 of the next higher level of recursion.

System 2—System 2 deals with damping oscillations among the operational units of the System 1 parts. Oscillations derive from the interaction of the operational units in the vertical domain (zigzag line shown in Figure 1). Therefore, System 2 is required to handle such disturbances among the operational units.

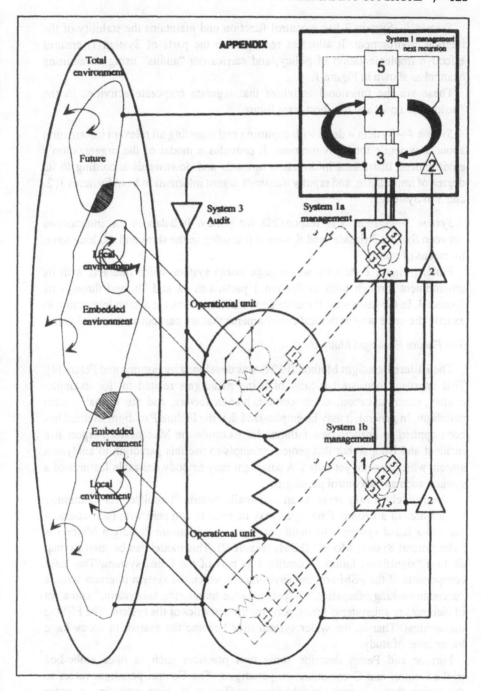

Figure 1. The Viable System Model (VSM).

System 3—System 3 has a control function and maintains the stability of the internal environment. It allocates resources to the parts of System 1, ensures effective implementation of policy, and carries out "audits" using an auditing channel as shown in Figure 1.

These are the functional activities that regulate corporate activities in the "inside and now" and the short-term future.

System 4—System 4 deals with capturing and reporting all relevant information about a system's total environment. It provides a model of the organization's environment, distributes information upwards and downwards according to its degree of importance, and rapidly transmits urgent information from Systems 1, 2, and 3 to System 5.

System 5—System 5 is responsible for policy. It balances the interactions between System 3 (inside) and System 4 (outside), in the short and the long term, by means of norms.

Finally, Figure 1 shows a whole page viable system which interacts with its environment through both its System 1 parts, i.e., 1a and 1b, and through its System 4. In the same way, the embedded viable systems are shown interacting in exactly the same way with local environments that are particular to each of them.

The Failure Paradigm Method

The Failure Paradigm Method (FPM) was developed by Fortune and Peters [9]. This approach consists basically of using paradigms related to, for example: control, communication, safety culture, human factors, and the formal system paradigm. In general, it may be emphasized that the Failure Paradigm Method has been applied to analyze past failures, for example the Manchester Airport fire incident and Bhopal, but this project attempts to use this paradigm to analyze a system which has not yet failed. A paradigm may embody desirable features of a system, such as the control paradigm.

Alternatively it may serve as an essentially neutral "template" for examining the features of a system. Paradigms may be used to compare with their counterparts in a failed system. The main paradigm of the Failure Paradigm Method is "The Formal System Model" (FSM) (Figure 2). This model can be used to map on to a "significant failure," identified as part of the failed system. The main components of the FSM are: the environment which the system interacts with, a "decision-making subsystem," a "performance monitoring subsystem," and a set of elements or subsystems which perform the activities of the system. The FSM is hierarchical. That is, the wider system may become the system in focus for a lower level of study.

Fortune and Peters describe some best practices such as ones embodied in the Control and Communication paradigms. The Control paradigm refers to the action that a system or subsystem applies to its own activities in order to maintain a desired state. This paradigm employs three types of control

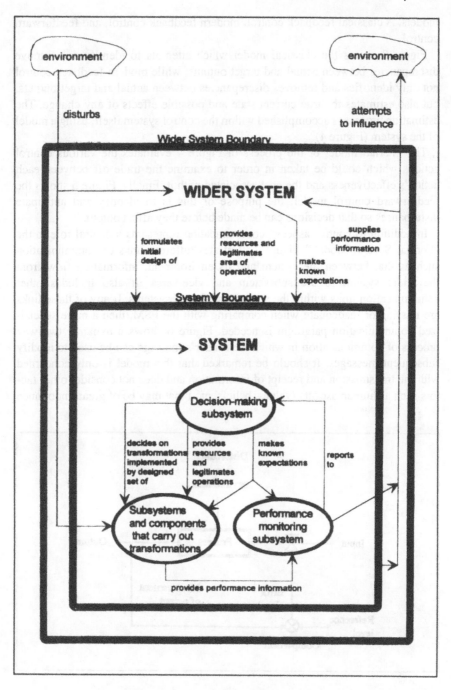

Figure 2. The Formal System Model [9].

models. A classical feedback control, modern feedback control, and feedforward control.

Figure 3 shows the classical model which attempts to identify and remove discrepancies between actual and target outputs; while modern feedback control not only identifies and removes discrepancies between actual and target outputs, but also estimates the true current state and possible effects of any change. The estimation process is accomplished within the control system itself, using a model of the system (Figure 4).

The internal model of the process of Figure 4 evaluates the various control actions which could be taken in order to examine the trade off between each action's effectiveness and the cost associated with it. Finally, Figure 5 shows the feedforward control model; the purpose of this is to identify and anticipate disturbances so that decisions can be made before they affect output.

In addition to control aspects, communication represents a central role in the "Formal System Model." Figure 2 illustrates different links of communication such as that between the system and its environment, information flow from the wider system to the subsystem and vice-versa. It also includes other communication links within the system and the subsystems. If any of these links are missing or inadequate when comparing with the FSM, then a more specialized communication paradigm is needed. Figure 6 shows a dynamic two-way process of communication in which the sender's message can be used to modify subsequent messages. It should be remarked that this model is only concerned with the transmission and receipt of information and does not consider other factors such as human aspects (values, beliefs, etc.) that may be of great importance

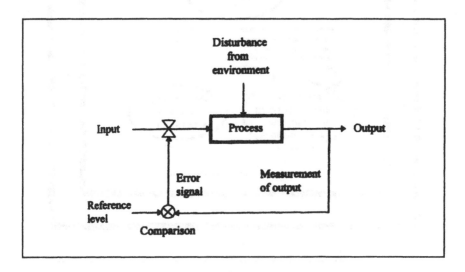

Figure 3. Classical feedback control.

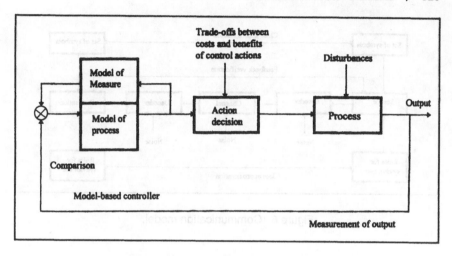

Figure 4. Modern feedback control.

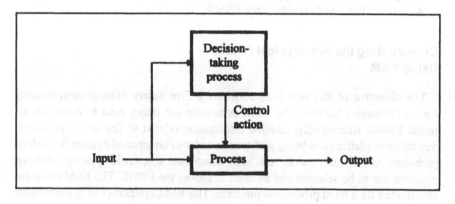

Figure 5. Feedforward control.

in the study of failures. To deal with this issue a communication model within and among teams is integrated in the methodology (FPM).

SYSTEMIC APPROACH TO FIRE SAFETY OFFSHORE

A four-step methodology is being used in an attempt to adopt a systemic approach to constructing a Fire Safety Management System (FSMS) for an off-shore platform. The stages are:

1. Construct a prototypical FSMS by using the Viable System Model.
2. Test the prototypical FSMS by applying the Failure Paradigm Method to it leading to a Synthesis in stage 3.

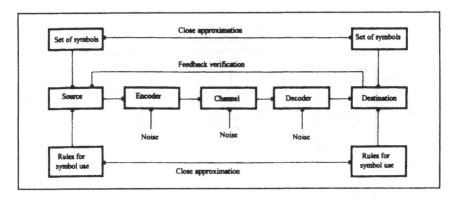

Figure 6. Communication model.

3. Synthesis, and
4. Further testing of aspects of the FSMS.

Constructing the Prototypical FSMS, Using VSM

The objective of this step is to construct a Fire Safety Management System for an Offshore platform. The VSM principles are being used to construct an initial FSMS. Accordingly, detailed information related to fire safety on North Sea offshore platforms is being gathered in order to understand current fire safety problems. Once this process has been completed a specific type of offshore platform has to be selected and be used to model the FSMS. The FSMS will be constructed for a fixed production platform. The VSM systems (1 to 5) mentioned above will then be identified and constructed.

Applying the Failure Paradigm Method

This step consists of comparing the initially formulated FSMS with the paradigms of the Failure Paradigm Method. Comparison between these paradigms and the formulated FSMS may produce discrepancies and deficiencies. From this comparison changes and improvements may be achieved.

Synthesis

The aim of this stage is to reassess the FSMS prototype according to the weaknesses detected in the previous stage. An effective FSMS is expected to be the final result of this step.

Further Testing of Aspects of the FSMS

Basically, a FSMS could only really be tested through its use in the real world, over a number of years. However, it may be possible to test aspects of the FSMS. For example, introducing an external disturbance requiring an emergency response. It is intended to conduct computer-based simulations of such an emergency in order to shed light on the effectiveness of the FSMS.

CONCLUSION

A new way of looking at fire risk on an offshore platform is described. This way of looking at fire safety is an alternative to earlier thinking and it is hoped that it will prove itself more satisfactory for understanding the nature of fire risk. A FSMS model is being constructed using the concepts of systems. This new way of looking at fire risk may be applied in a similar way to safety in general and to any organization. Ultimately, it is hoped that it will lead to more effective management of fire safety offshore.

REFERENCES

1. A. N. Beard, Towards a Systemic Approach to Fire Safety, *First International Symposium on Fire Safety Science,* October 7-11, 1985, Washington, D.C.
2. A. N. Beard, Some Ideas on a Systemic Approach, *Fire Safety Journal, 14,* pp. 193-197, 1989.
3. The Hon. Lord Cullen, *The Public Inquiry Into Piper Alpha Disaster,* Department of Energy, HMSO, 1990.
4. B. Weibye, Safety Management Under Different Regimes, *Offshore Northern Seas (ONS) Conference,* August 27-30, 1996, Stavanger, Norway.
5. S. Beer, *Decision and Control—The Meaning of Operational Research and Management Cybernetics,* John Wiley, Chichester, 1995.
6. R. L. Ackoff, *The Second Industrial Revolution,* manuscript, Wharton School of Finance and Commerce, University of Pennsylvania.
7. S. Beer, *The Heart of Enterprise,* John Wiley, Chichester, 1979.
8. S. Beer, *Diagnosing the System for Organisations,* John Wiley, Chichester, 1985.
9. J. Fortune and G. Peters, *Learning from Failure—The Systems Approach,* John Wiley, Chichester, 1995.

CHAPTER 10

Salt Water Simulation of the Fire Smoke Movement in an Atrium Building

H. P. Zhang, W. C. Fan, T. J. Shields,
and G. W. H. Silcock

Salt water simulation was used to study the movement characteristics of smoke in an atrium building. The flow patterns in an atrium were simulated using a salt water modeling technique. Both the buoyant plume and circulation patterns were observed, and the mechanism of the formation of the circulation was analyzed. Based on the analysis of the experimental results, the velocity of the salt water plume was deduced together with criteria to determine the initial direction of the salt water circulation plume. The maximum atrium height at which circulation does not occur was estimated by studying the effects of the enclosing structure and the source energy on the flow pattern in the atrium.

Recent decades have seen a worldwide increase in construction of atrium buildings and buildings containing atriums. Usually atrium buildings are very tall and occasionally are characterized by complex internal geometry, such that some fire engineering technologies are rendered inappropriate. Studies to date have included experimental determination of smoke evacuation in a large atrium [2], smoke control and venting [3, 4], modelling smoke movement, empirical evaluation of design parameters, and numerical simulation [5-7]. Although a design code BS 5588-PT 7 1997 Fire Precautions in the Design, Construction and Use of Buildings; Part 7 Code of Practice for the Incorporation of Atria in Buildings has just been published in the United Kingdom, much of the research work related to fire and atriums may be considered to be at a preliminary phase.

In atria, the process of the smoke spread from a corridor to an atrium is a transformation process which causes a vertical buoyancy plume to change into a corridor horizontal ceiling jet then to a vertical atrium buoyancy plume. In the single room, the momentum and kinetic energy flux are mainly induced by the pressure difference between the zone around the single room ceiling and the atrium. Generally speaking, the induced buoyancy plume caused by the corridor jet after it enters the atrium is 3-dimensional and unsteady flow. While the smoke plume moves upwards in the atrium, heat and mass transfers occur between the plume and the ambient air. Until now, it is still very difficult to describe the process accurately in a mathematical form since the complex structure and huge size of an atrium make it difficult to conduct full scale tests. Therefore, a simulation study is still one of the most effective and economical methods by which to research fire smoke movement in an atrium.

Although salt water simulation has been long recognized as one of the most effective means for the study of building fire smoke movement, it has seldom been applied to the study of the behavior of smoke in atria. In most research conducted to date, it has been assumed that there are enough venting orifices to provide the plume in an atrium with sufficient ambient air. Thus, the smoke plume in a corridor-atrium configuration is often simplified as a 2-D linear plume. The simplification is reasonable under certain conditions, but often the case where there is only one vertical inlet and one horizontal outlet and inlet vent is from the adjoining single room and/or corridor is ignored. Salt water simulation of the smoke movement in an atrium under these particular conditions was conducted and the procedures adopted and the results obtained are discussed in this chapter.

APPARATUS AND PROCEDURE

Apparatus and Room-Corridor-Atrium Model Design

As shown in Figure 1, a 1/20 scale room-corridor-atrium was immersed upside down in a fresh water tank. The details of the structure of the model are as shown in Figure 2. The atrium had an open top and a low level vent. The dimensions of the room, corridor, and atrium were: 0.8 m × 0.065 m × 0.08 m; 0.38 m × 0.05 m × 0.075 m; and 0.13 m × 0.20 m × 0.40 m, respectively. The dimensions of the doors between the room and corridor and atrium were 0.10 m × 0.05 m. The salt water was introduced at low level at center of the room through a nozzle 0.02 m in diameter. As in previous studies [8], the flow of salt water through the nozzle simulates the heat output in the room.

Scaling Analysis

As established in other related work [8], provided the geometrical similarity and other dynamic similarity criteria are satisfied and that chemical reactions are

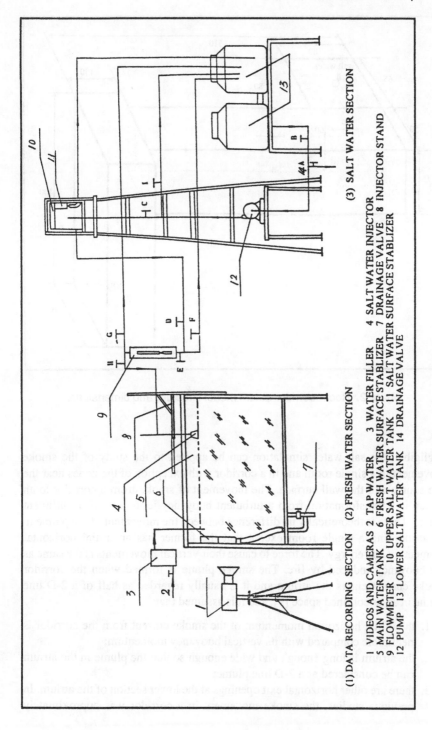

(1) DATA RECORDING SECTION   (2) FRESH WATER SECTION

(3) SALT WATER SECTION

1 VIDEOS AND CAMERAS 2 TAP WATER   3 WATER FILLER   4 SALT WATER INJECTOR
5 MAIN WATER TANK   6 FRESH WATER SURFACE STABLIZER   7 DRAINAGE VALVE   8 INJECTOR STAND
9 FLOWMETER   10 UPPER SALT WATER TANK   11 SALT WATER SURFACE STABLIZER
12 PUMP   13 LOWER SALT WATER TANK   14 DRAINAGE VALVE

Figure 1. Experiment apparatus.

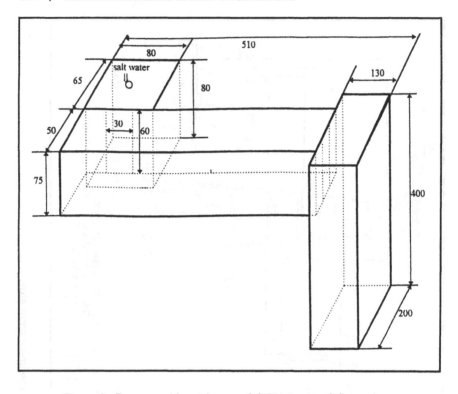

Figure 2. Room-corridor-atrium model structure and dimensions.

negligible, the salt water simulation can be applied to the study of the smoke movement in a single room and in a corridor with exception of the zones near the fire source and the wall surface. The movement of smoke from a corridor to an atrium is a transformation from a turbulent buoyancy jet to a vertical turbulent plume. There is theoretically no difference between the movement of the plume in an atrium and a single room except that the former has an initial horizontal momentum and energy. The force to cause their vertical movements is the same as the buoyancy induced by fire. The smoke plume is formed when the corridor smoke current enters the atrium and it is usually regarded as half of a 2-D line plume in an unconfined space [9], and it is assumed that:

1. the initial horizontal momentum of the smoke current from the corridor is negligible compared with its vertical buoyancy momentum;
2. the atrium is long enough and wide enough so that the plume in the atrium can be considered as a 2-D line plume;
3. there are other horizontal exit openings at the lower section of the atrium. In previous studies, the smoke movement in a corridor was approximately

regarded as a 2-D flow given that the corridor length is not less than twice the corridor height (which is the case for the model used in this study). The flow is not affected by the geometry of the fire source room or the way the smoke source induces the ambient air, i.e., the most important factor to affect the flow is the source strength. For smoke movement in a room- corridor-atrium building, the fire in the room can be considered as a point fire source, the smoke flow in the corridor can be regarded as a stratified layer, and the smoke movement from the corridor to the atrium can be approximated as a 2-D line plume formed at the exit of the corridor.

Experimental Results Observation

The experimental study was conducted using the apparatus as shown in Figure 1. Salt water was injected at the center of the room floor at low level, at mass rates in the range of 0.005 kg/s to 0.020 kg/s for relative densities of 0.02, 0.05, and 0.10, respectively. The relations between the resulting flow patterns of salt water current in the atrium, time, salt water mass flux, and relative density were obtained.

It was noted that:

1. The salt water current in the atrium is not merely a simple plume near to the wall. This is due to the fact that when the salt water current flows from the corridor into the atrium where it forms a plume, the axis line is located away from the wall surface of the atrium. This is due to the effect of the initial momentum of the salt water current, Figure 2. It has been observed that the deviation of the plume increases with the increasing of the height of the plume in the atrium. When the front of the plume reaches a certain height in the atrium, part of the plume will gradually turn around, down instead of rising, and form a large circulation plume in the XOZ plane of the atrium, Figures 3 and 4.

2. The rotational direction of the circulation in the atrium is constantly changing, Figures 3 and 4.

3. The initial rotational direction of the circulation is determined by the momentum of the salt water current in the corridor, i.e., the greater the momentum, the more likely the initial rotational direction is clockwise. Also, the frequency of the change of the rotational direction is also related to the normalized density difference Δ, i.e., $(\rho-\rho_0)/\rho_0$ and mass flow rate $\dot{m}_0$ of the current, i.e., the frequency decreases when Δ increases and $\dot{m}_0$ decreases.

From the observations and results obtained, it is apparent that the flow patterns of smoke in the atrium are more complex than that for a single room and a corridor configuration. The measurement methodologies and theoretical analysis currently

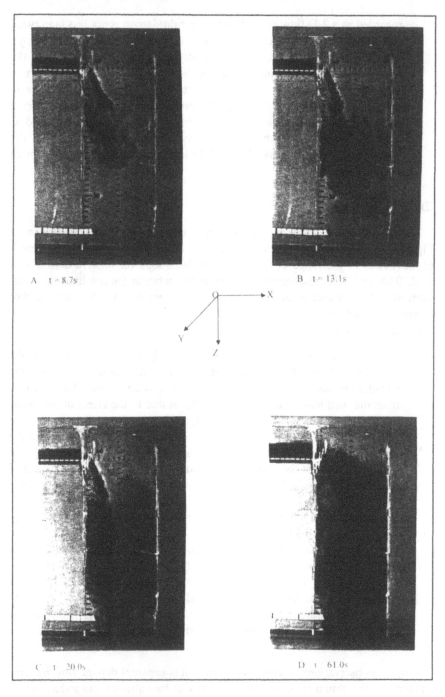

Figure 3. Atrium salt water flow pattern maps ($\Delta = 0.02$, $\dot{m}_0 = 0.005$ kg/s).

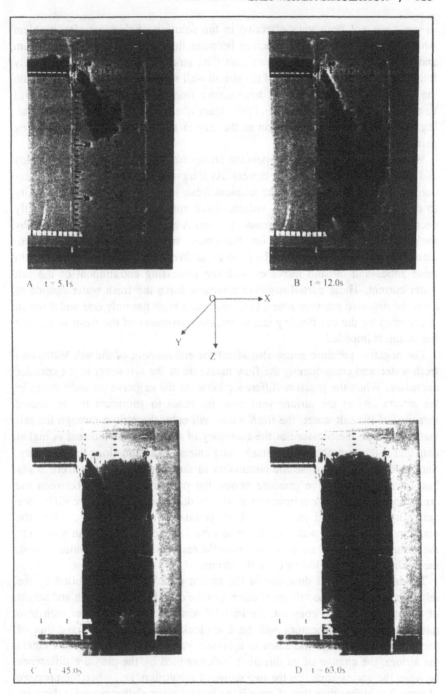

Figure 4. Atrium salt water flow pattern maps ($\Delta = 0.02$, $\dot{m}_0 = 0.009$ kg/s).

in use have not been very effective in the solution of this complex problem which are caused by the interaction between the smoke plume in the atrium and the air available at the atrium vent (i.e., stratified currents are not usually found in an atrium). The effect of the atrium wall surfaces will vary depending on the size of the atrium and the heat output from the fire. Further study will be required to solve this problem. Preliminary qualitative analysis of the generating mechanism of the circulation in the atrium are presented in the following paragraphs.

When the salt water current enters the atrium from the corridor, only buoyancy and gravity forces influence the current. As it travels in the atrium, the salt water current continuously entrains the ambient fresh water which causes its density to decrease. Consequently, the volume force applied to the current gradually decreases. Vortex and negative pressure zones A and B, Figure 5, are formed by the entrainment of the salt water into fresh water which is caused by the confinement imposed by atrium walls. The pressure difference between the fresh water in these zones and outside increases with the increasing entrainment of the salt water current. These partial negative pressures force the fresh water outside to enter the negative pressure zones, but, since the atrium has only one outlet which is occupied by the out flowing salt water, the movement of the fresh water into the atrium is impeded.

The negative pressure zones also affect the entrainment of the salt water into fresh water, and consequently the flow mass rate of the salt water in the corridor decreases. When the pressure difference between the negative pressure zones in the atrium and at the atrium vent zone increases to surmount the resistance (gravity) of the salt water, the fresh water will enter the atrium through the salt water. Because the pressure at the boundary of the salt water current is higher than that at the central zone, the fresh water enters the atrium along the boundary. Such two-way flows cause the circulation in the atrium. Once the fresh water has entered the negative pressure zones, the pressure difference between the zones and the ambient environment gradually decreases, the pressure difference between the zones and the source room gradually increases, and so does the entrainment of the salt water source. When the force to drive the fresh water into the atrium becomes too small to overcome the resistance of the salt water current, the fresh water will no longer enter the atrium.

The initial rotational direction of the atrium circulation is controlled by the initial momentum of the salt water current in the corridor. If the length and height of the atrium are kept constant, the initial direction of the salt water with zero initial horizontal momentum will be anti-clockwise. When the mass flux of the injecting salt water increases to a certain value, the initial direction enters the atrium, the change of its direction is determined by the pressure difference between the vortex zones on the two wings of the atrium, i.e., when the pressure of zone A is higher than that of zone B, a clockwise circulation occurs, otherwise an anti-clockwise circulation will occur.

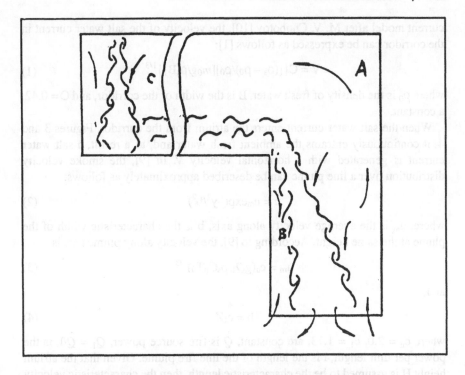

Figure 5. Circulation plume generation mechanism.

ANALYSIS OF SALT WATER MODELING TO PREDICT
ATRIUM FIRE SMOKE MOVEMENT

From analysis of the experimental results, a few distinct phases of smoke movement in the atrium are apparent:

1. the smoke plume and ceiling jet in the fire source room,
2. the horizontal stratified smoke layer in the corridor, and
3. the smoke overflow from corridor to atrium and the circulation plume in the atrium.

Conventionally, smoke spread modeling techniques for single room and corridor can be applied in the similar spaces adjacent to the atrium, thus only the plume velocity, circulation direction, with respect to height in the atrium, are discussed.

Atrium Plume Velocity

If salt water with a density ρ_s and an injecting mass flow rate of $\dot{m}_0$ is used to simulate the fire source power in a room, then, by applying the horizontal passage current model after M. V. Chobotov [10], the velocity of the salt water current in the corridor can be expressed as follows [1]:

$$V = C\{[(\rho_s - \rho_0)/\rho_0][\dot{m}_0 g/\rho_s B]\}^{1/3} \tag{1}$$

where ρ_0 is the density of fresh water, B is the width of the corridor, and $C = 0.42$, a constant.

When the salt water current enters the atrium from the corridor, Figures 3 and 4, it continuously entrains the ambient fresh water and, as a result, a salt water current is generated with a horizontal velocity V. In [9], the smoke velocity distribution over a line plume can be described approximately as follows:

$$u = u_m \exp(-y^2/b^2) \tag{2}$$

where, u_m is the average velocity along axis, b is the characteristic width of the plume at the same height. According to [9], the velocity along plume axis is

$$u_m = c_u (g\dot{Q}_1/\rho_s C_p T_0)^{1/3} \tag{3}$$

and;

$$b = c_1 Z \tag{4}$$

where $c_u = 2.0$, $c_1 = 1.13$, are constant, $\dot{Q}$ is fire source power, $\dot{Q}_1 = \dot{Q}/l$, is the power per unit length, l is the length of the line fire plume. Given that the atrium height H is assumed to be the characteristic length, then the characteristic velocity is expressed as for the smoke plume in the atrium.

$$u_0 = (\dot{Q} g/\rho_s C_p T_0 H)^{1/3} \tag{5}$$

From equations (3) and (5), the non-dimensional velocities $\hat{u}_m$ and $\hat{u}$ can be written as,

$$\hat{u}_m = c_u (l/H)^{-1/3} \tag{6}$$

and

$$\hat{u} = \hat{u}_m \exp(-y^2/b^2)$$
$$= c_u (l/H)^{-1/3} \exp(-y^2/b^2) \tag{7}$$

According to scaling theory, the non-dimensional velocity $(\hat{V}_z)$ of the salt water at Z-axis should be expressed by

$$\hat{V}_z = c_u (B/H)^{-1/3} \exp(-y^2/b^2) \tag{8}$$

therefore, the average velocity V_z of the salt water is

$$V_z = 2.0(\dot{m}_0 g/\rho_s B)^{1/3} \exp(-y^2/b^2) \tag{9}$$

The horizontal velocity of the salt water in the atrium is different from that in the corridor because the salt water entrains ambient fresh water while it spreads in the atrium, the mass rate of the salt water increases. According to the momentum principle, the horizontal velocity of salt water will decrease gradually. From equation (9), the entrainment of the salt water into fresh water can be given,

$$\dot{m}_{es} = \int_0^\infty 2BV_z\rho_s dy$$

$$= \int_0^\infty 4.0B\rho_s \, (\dot{m}_0 \, g/\rho_s B)^{1/3} \exp(-y^2/b^2)$$

or,

$$\dot{m}_{es} = 0.51 \, (\dot{m}_0 \, g)^{1/3} \, (\rho_s B)^{2/3} \tag{10}$$

Furthermore, the horizontal velocity of the salt water V_x is,

$$V_x = \dot{m}_0 \, V/(\dot{m}_0 + \dot{m}_{es}) \tag{11}$$

or,

$$V_x = \{C\dot{m}_0/(\dot{m}_0 + \dot{m}_{es})\} \cdot \{[(\rho_s - \rho_0)/\rho_0][\dot{m}_0 \, g/\rho_s B]\}^{1/3} \tag{12}$$

Initial Rotational Direction and Frequency of Circulation

From observation it was clear that a circulation plume was formed when a salt water current entered the atrium under certain conditions, i.e., when $H/L \geq 3$, and the direction of the resulting circulation was changing with time. It is difficult to quantify the factors which induce such changes to the direction of the circulation currents. However, it has been established by experiment that the initial direction of the circulation is non-stochastic (non-random) and can be determined by the initial momentum of the corridor salt water current and by the buoyancy plume momentum of the current at the inlet to the atrium:

$$\text{i.e., } E = \rho_s V^2/[2(\rho_s - \rho_0)gH]^{1/3} \tag{13}$$

Apply equation (1) to (13), we get,

$$E = C^2(\Delta \dot{m}_0/\rho_s B)^{2/3}/[2(\rho_0/\rho_s)\Delta gH] \tag{14}$$

assuming $\rho_0 \approx \rho_s$, equation (14) can be rewritten as:

$$E = C^2 \, (\dot{m}_0/\rho_s)^{2/3}/[2(\Delta g)^{1/3}B^{2/3} \, H] \tag{15}$$

As shown in Table 1, the initial direction of the atrium circulation plume is affected by the value of E. Therefore, the value can be used as a criteria to decide the initial direction of the circulation plume in the atrium to some degree.

$$E \geq 1.5 \times 10^{-3}, \text{clockwise circulation;} \tag{16}$$

$$1.5 \times 10^{-3} > E > 1.2 \times 10^{-3}, \text{difficult to decide;} \tag{17}$$

$$1.2 \times 10^{-3} \geq E, \text{anti-clockwise circulation.} \tag{18}$$

The frequency of the circulation direction was not regular and only the mean results obtained are given in Table 2. From the results obtained it is clear that the frequency increases approximately as m_0 increases, Δ decreases and that with changes in Δ between 0.02 and 0.1, m_0 between 0.005 kg/s and 0.018 kg/s, this frequency is between 1.0×10^{-2}/s ~ 4.0×10^{-2}/s.

Atrium Circulation Critical Height

In an atrium the circulation of smoke within the atrium usually will affect the smoke drainage potential from the atrium. The height of the atrium is critical, i.e., the higher the atrium the more likely the circulation of the smoke. Beneath a certain height no circulation will occur, this height is deemed the atrium circulation critical height.

Table 1. Atrium Circulation Plume Initial Direction versus E

$E (\times 10^{-4})$	4.0	6.0	8.0	9.0	10.0
Circulation direction	anti-clockwise	anti-clockwise	anti-clockwise	anti-clockwise	anti-clockwise
$E (\times 10^{-4})$	12.0	13.0	15.0	17.0	20.0
Circulation direction	anti-clockwise	difficult to decide	clockwise	clockwise	clockwise

Table 2. Average Circulation Direction Changing Frequency versus m_0 and Δ

f(l/s) Δ \ m_0	0.005 (kg/s)	0.010 kg/s	0.015 kg/s	0.020 kg/s
0.02	0.013	0.015	0.023	0.034
0.05	0.011	0.014	0.018	0.033
0.10	0.010	0.016	0.021	0.029

Generally, the inducement of the circulation flow is due to the supply air which cannot smoothly enter the negative pressure zones. Consequently, it can be assumed that the extremities of the salt water band increases as the distance moved by the plume in the atrium increases. When this distance increases to a certain value, the plume encompasses the vent outlet and the height at which this occurs is deemed the atrium circulation critical height.

From experimental observations, there appears to be basically two distinct patterns of flows occurring in the atrium.

1. The salt water current is an upside down plume close to the wall surface, as shown in Figure 6. As the plume is half of the line plume in a non-confined space, the condition for the circulation not to occur is that the half of the width of the plume is no more than the length (L) of the atrium.
2. When salt water current moves in the atrium, both the axis line and the boundaries of the current deviate from the wall surface because of the horizontal momentum of the current. When the current reaches a certain height in the atrium, the boundaries far from the inflow into the plume will encounter the side wall of the atrium and circulation will occur as a result. The condition for the circulation not to occur is that the atrium height is less than the distance from the corridor vent to the point where the salt water current plume boundary meets the atrium wall surface as shown in Figure 7.

Consider the flow patterns illustrated in Figure 6, if the width of a non-confined line plume at the height Z is 2b, then the relation between b and Z is [9]:

$$2b = 1.13Z \tag{19}$$

when the height of the atrium is H, the height of the corridor is h, $H = Z + h$, the condition for circulation not to occur is:

$$b < L \tag{20}$$

then

$$Z < 2L/1.13 \approx 1.8L \tag{21}$$

Therefore, the maximum atrium height H_{lmax} for the circulation not to occur is:

$$H_{lmax} \approx h + 1.8L \tag{22}$$

or

$$H_{lmax}/L \approx h/L + 1.8 \tag{22a}$$

The flow pattern in the second case is much more complicated. So, for simplification, we assume that the salt water in the atrium is a line plume deviating from the wall and the plume has movement both in X and Z directions, as

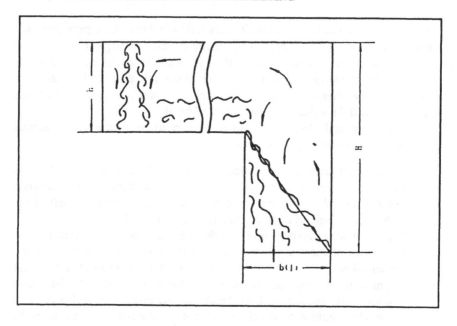

Figure 6. Salt water flows structure analysis (a).

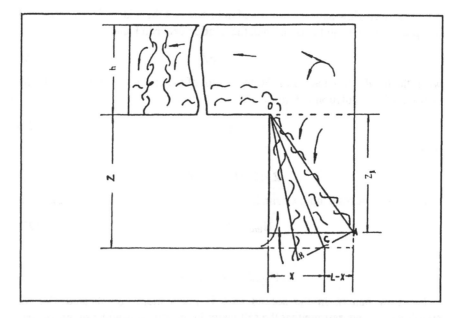

Figure 7. Salt water flows structure analysis (b).

shown in Figure 7. The velocities in X and Z directions can be described by equations (9) and (12), respectively.

Suppose H_{2max} is the maximum atrium height for the circulation not to occur, b is the half width of the plume, X is the maximum deviation of the salt water current from the axis line, and L is the length of the atrium. Taking the average value of the initial velocity as it enters the atrium and the final velocity when it reaches the wall of the atrium as the horizontal velocity of the plume, the following equations can be generated,

$$b = 1.13Z/2 \tag{23}$$

$$X = \overline{V} \times t = \overline{V} \quad Z/V_z|_{y=0}$$

$$= \{(2\dot{m}_0 + \dot{m}_{es})/2(\dot{m}_0 + \dot{m}_{es})\}(C \, \Delta^{1/3}/2.0)Z \tag{24}$$

$$Z_1 = (Z^2 + X^2 + b^2 - L^2)^{1/2} \tag{25}$$

$$Z_1 = Z - \{(b^2 - (L - X)^2\}^{1/2} \tag{26}$$

$$H_{2max} = Z_1 + h \tag{27}$$

or

$$H_{2max}/L = Z_1/L + h/L \tag{27a}$$

From equations set (23) to (27), we can obtain the value of H_{2max}, the maximum atrium height in the second case. Both cases can appear in the development of salt water flows in the atrium; therefore, the maximum atrium height not to generate a circulation under the experimental condition mentioned above is:

$$H_{max} = min \{ H_{1max}, H_{2max}\} \tag{28}$$

What must be pointed out is that equation (28) is only an estimation equation based on experimental observations and measurement. Therefore, the effect of the false line source was not taken into consideration in the deduction. Otherwise, the value of H_{2max} should be expressed as $H'_{2max} = Z' + h - Z_0$, where Z_0 is the distance from false line source to line source, Figure 8.

Figure 9 describes the effect of the mass flux rate (m_0) of the injected salt water and the relative density Δ on the rate of H_{max} to atrium length L. As shown in the experimental results, H_{max}/L decreases along with the increasing of Δ and m_0 with m_0 effect being much stronger. Under the experimental conditions, the values range of H_{max}/L were between 1.7 and 2.6 which was in accord with the estimation given by equation (22). Therefore, in order to increase H_{max}, L must be increased and Δ and m_0 both decreased. Practically, the most effective measure is to make fresh water enter at the bottom or lower section of the atrium so that the induced fresh water no longer enters the atrium from the opening and thus avoids

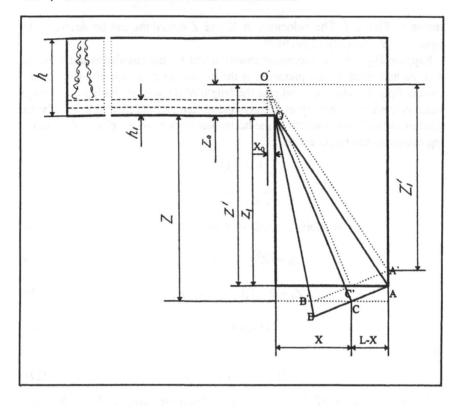

Figure 8. Effect of false line source position on H_{max}.

the generating of the two way currents. Since there are usually vents at the lower parts of a real atrium, the circulation currents seldom appear in a real atrium. In brief, the atrium discussed in the study is one with complicated smoke movement and poor efficiency of smoke evacuation, i.e., condition to be avoided in the design of atria building.

CONCLUSIONS

From the observations and analysis of experimental results obtained in salt water simulation of smoke movement, it follows that:

1. Salt water simulation can be applied in the study of the fire smoke movement not only in a single room and a corridor, but also in an atrium.

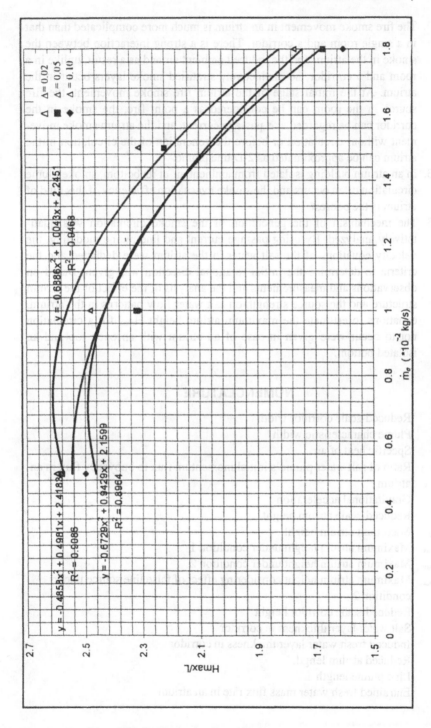

Figure 9. Hmax/L versus $\dot{m}_o$.

2. The fire smoke movement in an atrium is much more complicated than that in a single room and a corridor. There is a strong interaction between the smoke in the atrium and the induced ambient air and as a result, unlike in a room and a corridor, no distinctively stratified smoke layers occur in the atrium. At the initial stage of the atrium fire smoke movement, the fire source in the room can be considered as a point fire, the smoke in the corridor can be regarded as a passage current, and the atrium smoke movement without circulation or before the generation of the circulation in the atrium can be approximated as 2-D line plume.

3. In an atrium building isolated from ambient air at its bottom, smoke plume circulation often occurs and the smoke evacuation efficiency of this kind of atrium is the poorest.

3. The mechanism of the generation of the smoke circulation was qualitatively analyzed. By using passage current and line plume modeling, the velocity equations of the salt water in the atrium were deducted and the criteria to determine the initial rotational direction was given. Based on observations and measurements and the analysis of the effect of the atrium structure and the source strength on salt water flow patterns in the atrium, equations to estimate the maximum atrium height H_{max} for a circulation not to occur were given in case of an atrium with an open top and an isolated bottom.

NOMENCLATURE

B	Reduced-scale corridor width
b	Plume characteristic width
C_p	Specific heat of air
E	Ratio of salt water plume's initial momentum flux to buoyancy flux in an atrium
g	Gravitational acceleration
H	Reduced-scale atrium height
H_{max}	Maximum atrium height
H_{1max}	Maximum atrium height under condition 1
H_{2max}	Maximum atrium height under condition 2
H'_{2max}	Maximum atrium height considering effect of false line source under condition 2
h	Reduced-scale corridor height
h_1	Salt water layer thickness in corridor
h_2	Induced fresh water layer thickness in corridor
L	Reduced atrium length
l	Line plume length
$\dot{m}_{es}$	Entrained fresh water mass flux rate in an atrium

$\dot{m}_0$	Mass release rate of salt source
Q	Line fire source power
T_0	Ambient air temperature
t	Time
U_0	Characteristic velocity of smoke in air
u	Velocity distribution over a line plume
u_0	Characteristic velocity of smoke plume
u_m	Average plume velocity along Z-axis
$\hat{u}_m$	Dimensionless average plume velocity along Z-axis
V	Corridor average velocity of salt water current
V_x	Average salt water plume velocity in X-axis direction
V_z	Average salt plume velocity in Z-axis direction
$\hat{V}_z$	Dimensionless average salt water plume velocity
V_0	Characteristic velocity of salt water in fresh water
ρ	Density
ρ_0	Fresh water density
ρ_s	Salt water density
Δ	Normalized density difference $(\rho - \rho_0)/\rho_0$

REFERENCES

1. H. P. Zhang, *Salt Water Simulation of Fire Smoke Movement and the Study of the Movement Characteristics of Fire Smoke in A Confined Space*, doctoral dissertation, the University of Science and Technology of China, January 1996.
2. J. Tamegai, *Experimental Study on Performance on Smoke Control System by Pressurization in Central Tower*, Proceedings of the Annual Meeting Architectural Institute of Japan, Part A. Sendi, Japan, September 1991. (in Japanese)
3. J. H. Mcguire, G. T. Tamura, and A. G. Wilson, *Factors in Controlling Smoke in High Rise Buildings*, Symposium on Fire Hazards in Buildings, ASHRAE, San Francisco, 1970.
4. G. T. Tamura and C. Y. Shaw, *Experimental Studies of Mechanical Venting for Smoke Control in Tall Office Buildings*, ASHRAE Transactions, 86, 54, 1978.
5. W. K. Chow and W. K. Wong, On the Simulation of Atrium Fire Environment in Hong Kong Using Zone Models, *Journal of Fire Science, 11*:1, 1993.
6. W. K. Chow, Smoke Movement and Design of Smoke Control Atrium Buildings, *International Journal of Housing Science and its Applications, 13*, 1989.
7. W. K. Chow, Smoke Movement and Design of Smoke Control in Atrium Buildings—Part 2 Diagrams, *International Journal of Housing Science and its Applications, 14*, 1989.
8. H. R. Baum, J. G. Quintiere, and K. D. Steckler, *Salt Water Modeling of Fire Induced Flows in Multicompartment Enclosures*, NBSIR86-3327, U.S. Dept. of Commerce, National Bureau of Standards, National Engineering Laboratory, Center for Fire Research, Maryland, March 1986.

9. Y. Liming and G. Cox, Measurement of the Mass Flux of 2-D Line Fires Plume, *Fire Safety Journal*, 4:3, 1995. in Chinese.
10. M. V. Chobotov, E. E. Zukoski, and T. Kubota, *Gravity Current Heat Transfer Effects, NBS-GRC-87-522*, California Institute of Technology, Pasadena, California, December 1986.

CHAPTER 11

Salt Water Simulation of the Movement Characteristics of Smoke and Induced Air in a Room-Corridor Building

H. P. Zhang, W. C. Fan, T. J. Shields,
and G. W. H. Silcock

A salt water simulation and the double-liquid-dyeing technique have been applied to study the characteristic movement of smoke and induced air in a corridor adjoining a room. The occurrence of stratified smoke and air layers in the corridor was verified and the interaction between the two layers determined to be negligible. The position and heat release rate of the fire source in the room of fire origin affect the smoke layer and the induced air layer velocities but do not affect their corresponding non-dimensional velocities. By the application of scaling theory and the analysis of experimental results, the single room point source plume model, the horizontal passage gravity currents model, and equations concerning the movement of the smoke and the induced air were derived. Such equations can be useful for fire smoke movement studies in a confined space.

A room-corridor arrangement is one of the most common features of buildings and plays a fundamental role in the spread of smoke during a fire. The toxic smoke and other combustion products generated during the early stage of a fire are known to spread very fast. As a consequence this leads to injuries and deaths of the occupants involved in a fire. During a fire, the corridor is an important passage for evacuation of occupants and for the fire-fighter access to fight the fire and rescue occupants. Thus it is important to have a good understanding of smoke movement and related human behavior if the hazard created by smoke and toxic products is to be reduced.

153

Due to the fact that the density of the hot smoke is lower than that of the air, a horizontal stratified smoke layer is formed under the corridor ceiling. Simultaneously, the entrainment of air into the fire plume in the fire source room causes cold air movement to the fire source room from the corridor, causing a reverse air flow in the lower part of the corridor. Thus air can potentially reach the fire source room, enter the smoke plume and the upper ceiling smoke layer. These air flows can also affect the development of fire smoke spread and also alter the ventilation characteristics of the fire zone [1, 2]. Therefore, a study of the movement of smoke and accompanying induced air flow in a corridor is both timely and appropriate.

A problem worthy of attention involves the spread velocity of a smoke layer, its thickness and the thickness of any induced air layer. Previous studies have been conducted to consider smoke movement from a constant heat release source located at one end of a long passageway. The behavior and characteristics of the spread of smoke from the single room at one end of the room-corridor building have been investigated [3, 4] and it was concluded that the room is either filled with smoke and air of different densities or is completely filled by smoke. But little attention was paid to how the movement of smoke and air in the corridor influenced the heat release rate in the fire source room. Few investigations to date have considered the smoke mass rate change induced by the flow around the fire source.

In this chapter, the fundamental mechanisms of smoke and entrained air flow and the method of salt water simulation [5] are discussed in detail. The movement characteristics of smoke and induced air were studied using a one-tenth scale room-corridor simulation model. By applying the double-liquid-dyeing technique, the stratified salt water and fresh water were visualized and their velocities measured. Finally, a smoke movement model and induced air flow model in a room-corridor are generated using scaling theory. An entrainment model of the plume in a single room and a gravity smoke current model in a passageway based on the experiment results were also developed. This is all based on the assumption that correlation derived for the analogous behavior of salt water and smoke are completely transferable.

SCALING AND ANALYTICAL STRATEGY

Scaling Strategy

It has been shown that the theoretical basis of studies of salt water simulation of fire generated smoke depends on two sets of dimensionless equations that describe the analogous movement of smoke in air and salt water diffusing in fresh water which are mathematically similar [5, 6]. The dimensionless equations describing smoke movement are as follows:

$$\frac{\partial \hat{u}}{\partial \hat{x}} = 0$$

$$\frac{\partial \hat{u}}{\partial \hat{t}} + \hat{u}\frac{\partial \hat{u}}{\partial \hat{x}} = -\frac{\partial \hat{p}'}{\partial \hat{x}} + \theta + \frac{1}{Re}\frac{\partial^2 \hat{u}}{\partial \hat{x}^2}$$

$$\frac{\partial \theta}{\partial \hat{t}} + \hat{u}\frac{\partial \theta}{\partial \hat{x}} = \frac{1}{Re\, Pr}\frac{\partial^2 \theta}{\partial \hat{x}^2} + G\hat{Q}_0$$

$$where,\ \theta = \frac{T - T_a}{T} \Big/ \frac{gH}{U_0^2},\ Fr^2 = \frac{gH}{U_0^2},\ G = (H/L)^3,$$

$$\hat{p}' = (p - p_a)\,/\,\rho U_0^2,\ U_0 = (\dot{Q}g\,/\,\rho_a C_p T H)^{1/3}$$

H is corridor height, T is smoke temperature, L is fire source diameter, p is pressures, U_0 is characteristic velocity of smoke, $\dot{Q}_0$ is fire source heat release rate, T_a, p_a, ρ_a represent temperature, pressure, and density of the ambient air, respectively.

For salt water spread, the equations are:

$$\frac{\partial \hat{u}}{\partial \hat{x}} = 0$$

$$\frac{\partial \hat{u}}{\partial \hat{t}} + \hat{u}\frac{\partial \hat{u}}{\partial \hat{x}} = -\frac{\partial \hat{p}'}{\partial \hat{x}} + \hat{F} + \frac{1}{Re}\frac{\partial^2 \hat{u}}{\partial \hat{x}^2}$$

$$\frac{\partial \hat{F}}{\partial \hat{t}} + \hat{u}\frac{\partial \hat{F}}{\partial \hat{x}} = \frac{1}{Re\, Sc}\frac{\partial^2 \hat{F}}{\partial \hat{x}^2} + G\hat{m}_0$$

$$where,\ \hat{F} = \frac{\rho - \rho_0}{\rho} \Big/ \frac{gh}{V_0^2},\ Fr^2 = \frac{gh}{V_0^2},\ G = (h/l)^3,$$

$$\hat{p}' = (p - p_0)\,/\,\rho_0 V_0^2,\ V_0 = (\dot{m}_0 g\,/\,\rho_0 h)^{1/3}$$

h is model height, ρ is salt water density, 1 is salt water source diameter, p is pressure, V_0 is characteristic velocity of salt water, $\dot{m}_0$ is salt water injecting mass rate, ρ_0 and p_0 represent fresh water density and pressure respectively.

Strictly speaking, we cannot get complete similarity of behavior between all of the dimensionless groups; however, reasonable results can be drawn from salt water simulation if the following conditions are satisfied.

1. The room-corridor model is geometrical similar to a full-scale building;
2. The fire source is relatively weak;

3. The Re number, Fr number, and the ratio of the characteristic length to fire source dimension, for both smoke movement and salt water spread, are numerically equal or nearly equal respectively; and

4. The initial momentum of the injected salt water can be ignored compared with its gravity momentum where the distance between the position and the inlet of salt water is one-tenth height of the model.

Thus the salt water simulation can yield correlations that predict behavior which can be then translated to issues of smoke and air flow movement.

Analytical Strategy

In this case it has been assumed, in addition to reduced scale modeling since salt water movement and mixing in fresh water is analogous to smoke in air, that any correlation between non-dimensional quantities that are found to exist for either medium are transferable and can be reduced to dimensional form. Also, it can be assumed that any existing correlations between dimensional parameters that have been validated in one medium can be modified to generate correlation in non-dimensional form which can in turn be validated for the other medium. Such an approach was used in the analysis of data as discussed in the following paragraphs.

APPARATUS AND PROCEDURE

As shown in Figure 1, the apparatus consists of three components: salt water system, fresh water system, and measurement system. As shown in Figure 2, a one-tenth scale room-corridor model is immersed upside down in the main fresh water tank. The dimensions of the model room are 0.32 m × 0.26 m × 0.30 m, the corridor 1.10 m × 0.20 m × 0.25 m, and the door 0.235 m × 0.115 m, the diameter of the entrance for salt water injection is 0.05 m.

The double-liquid-dyeing technique was used to visualize the movement of salt water and the fresh water in the room-corridor model at the same time and to determine velocities using video techniques so that the spread of the smoke and the opposite movement of the induced air could be studied. In this case, dyes of different colors from the salt water layer having the same density as the fresh water were injected into the fresh water. This allowed the dyes to move with the fresh water layer. It must be pointed out that the injecting velocity of the dyes should not exceed a velocity of 10^{-3} m/s in a direction normal to the fresh water flow, i.e., its influence on gravity momentum is neglible.

EXPERIMENTAL RESULTS AND ANALYSIS

Two behaviors and related results are shown in Figures 3 through 7. The movement velocities of the salt water and the fresh water in the corridor model

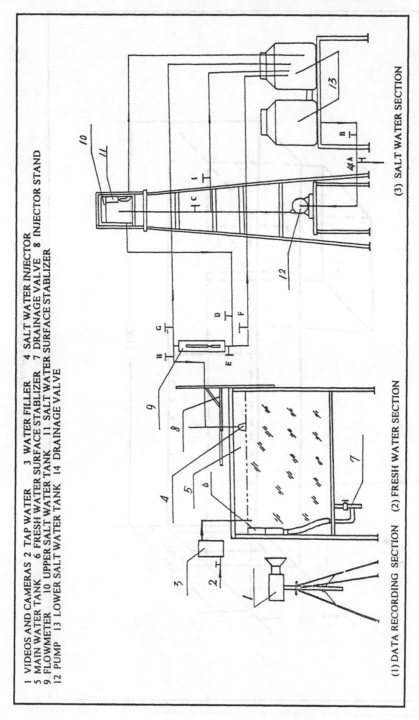

1 VIDEOS AND CAMERAS 2 TAP WATER   3 WATER FILLER   4 SALT WATER INJECTOR
5 MAIN WATER TANK 6 FRESH WATER SURFACE STABLIZER 7 DRAINAGE VALVE 8 INJECTOR STAND
9 FLOWMETER   10 UPPER SALT WATER TANK 11 SALT WATER SURFACE STABLIZER
12 PUMP 13 LOWER SALT WATER TANK 14 DRAINAGE VALVE

(1) DATA RECORDING SECTION   (2) FRESH WATER SECTION   (3) SALT WATER SECTION

Figure 1. Experiment apparatus.

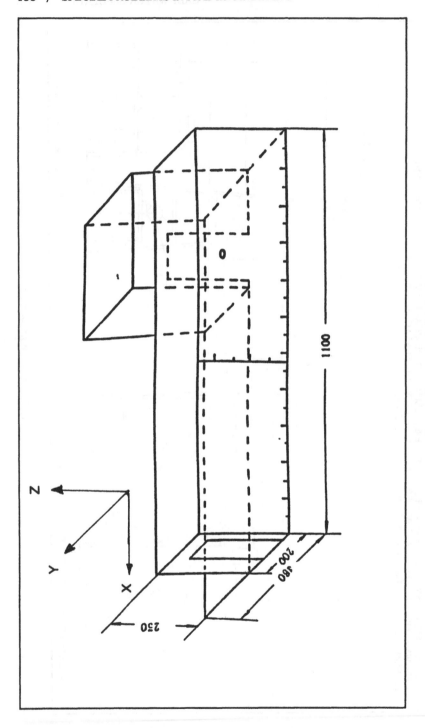

Figure 2. Room-corridor model.

were measured with variable salt water source positions, densities, and injection mass fluxes and the results obtained are illustrated. The other set of results obtained from the measurement of the thickness of the salt and fresh water layers are shown in Figure 8 and Figure 9. In all the experiments, the salt water was injected in the corner and in the center of the room. The densities of the salt water were 1100 kg/m^3, 1050 kg/m^3, 1020 kg/m^3, respectively, and the mass flux rate of salt water ranged from 0.005 kg/s to 0.038 kg/s.

From observation and analysis of the results, it was found that the characteristic movement and behavior of the salt water in a single room were similar to that of the smoke plume. When the stratified salt water layer under the ceiling reached the height of the door lintel, the salt water moved into the corridor through the door. Gradually a salt water layer of a certain thickness was formed under the ceiling. The layer spread outside and the thickness of the salt water layer remained constant along the length of the corridor, as shown in Figures 8 and 9. The fresh water moved along the corridor in the lower part of the corridor while the salt water layer reached the exit of the corridor. It was found that the velocities of both fresh water and salt water were constant as they moved in the corridor. It has been shown that the velocities are related to the mass flux of the injected salt water, the position of the injected salt water, and the normalized density difference of salt water $\Delta = (\rho - \rho_0)/\rho_0$. The simulation result indicates that the entrainment of the smoke into the ambient air in the corridor is negligible. As previously pointed out [5, 7], the mass flux, position of the injected salt water, and its normalized density difference affect entrainment. In other words, the velocities of the salt water and the contra fresh water flow in the corridor increase due to increasing of the mass flux, relative density, and the entrainment ability of the salt water source in the room. It should be noted that, as shown in Figure 7, the non-dimensional velocities of the salt water and the contra flow of fresh water do not depend on the mass flux rate. This is a very important observation from the salt water study which is in accord with the characteristics of the smoke movement ignoring any heat transfer heat losses to the walls.

It was also observed that the thickness of the salt water layer gradually increases with a corresponding decrease in the thickness of fresh water layer. After a period of time, the layer thicknesses achieved their quasi steady state layer equilibrium thickness. When this situation exists, the sum of the mass flux of the injected salt water and the induced entrainment equals the mass flux of the salt water layer flowing out of the door into the corridor. The main factors controlling the thickness of the salt water layer are the mass flux, position, and normalized density of the salt water. As shown in Figure 9, the equilibrium thickness increases as the mass flux rate increases. However, the equilibrium thickness drops as the normalized density increases as shown in Figure 8. Certainly, the equilibrium thickness is related to geometric conditions of the room exit of the corridor such as the height and width of the door, as well as the distance between the lintel and the ceiling. Further studies on these problems will be conducted in the future.

ANALOGOUS VELOCITY MODELING OF SMOKE AND INDUCED AIR IN ROOM-CORRIDOR

Considering the heat release rate of the point fire source, $\dot{Q}_0$ kW as analogous to the salt water mass flux $\dot{m}_0$ kg/s, the movement behavior of the injected salt water in the room is similar to that of the smoke plume arising from a point fire

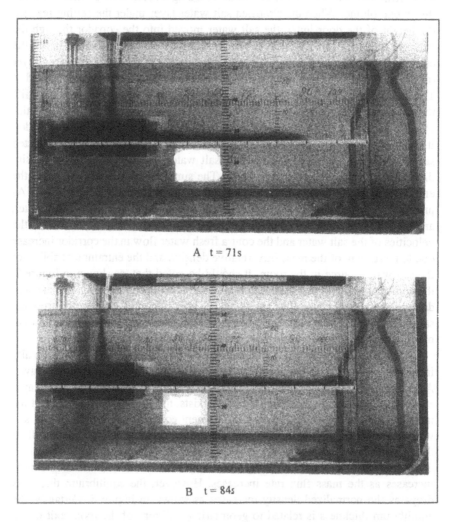

A t = 71s

B t = 84s

Figure 3. Measurement of velocities of salt water and induced fresh water in a corridor model (center source, $\dot{m}_0$ = 0.010/kg/s, Δ = 0.05).

source. Also, the movement of the salt water in the corridor is similar to that of the gravity air and smoke currents in a horizontal passageway. Therefore, the point source model for single room and the passage gravity current model can be used to analyze the movement of the salt water and the induced fresh water, and models concerning the smoke and induced air movement in a room-corridor building deduced.

Since, as previously argued, the mass flux of salt water out of the corridor ($\dot{m}$) equals the sum of the mass flux of the salt water injected into the room ($\dot{m}_0$) and

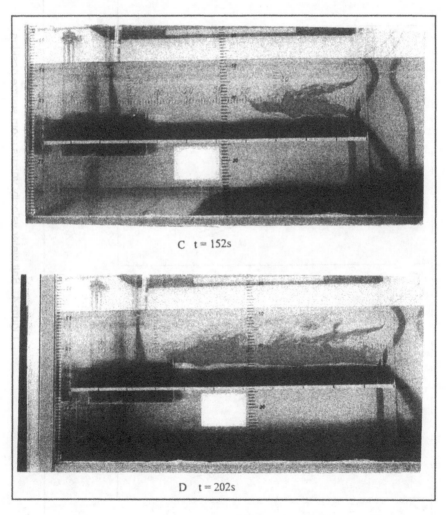

C t = 152s

D t = 202s

Figure 3. (Cont'd.)

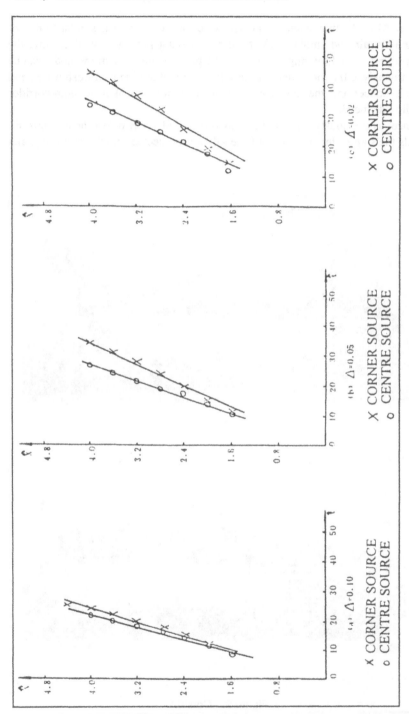

Figure 4. Salt water velocity versus room source position.

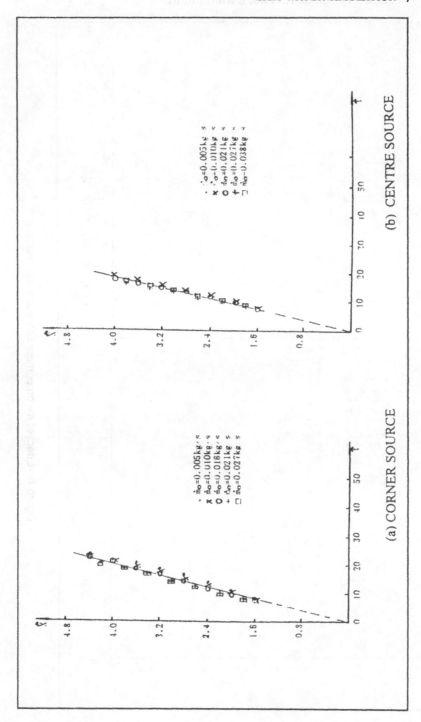

Figure 5. Corridor salt water flow velocity versus room source strength (Δ = 0.10).

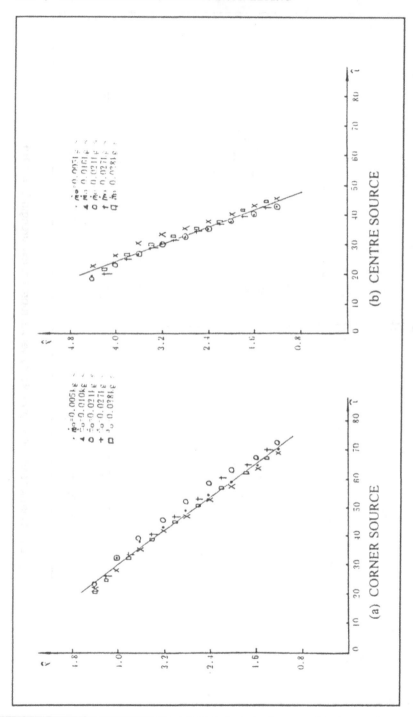

Figure 6. Effect of m_0 on corridor salt water flow velocity.

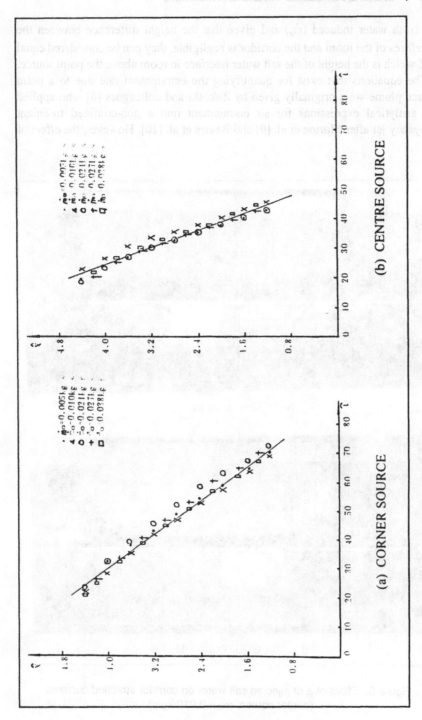

Figure 7. Effect of m_0 on corridor fresh water flow velocity.

the fresh water induced ($\dot{m}_e$) and given that the height difference between the interfaces of the room and the corridor is negligible, they can be considered equal to Z which is the height of the salt water interface in room above the point source.

The equations that exist for quantifying the entrainment rate due to a point source plume were originally given by Zukoski and colleagues [8] who applied the analytical expressions for air entrainment into a non-confined turbulent buoyancy jet after Morton et al. [9] and Baines et al. [10]. However, the effect of

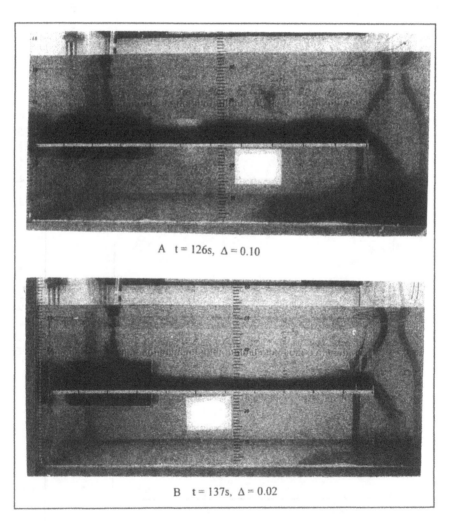

A t = 126s, $\Delta = 0.10$

B t = 137s, $\Delta = 0.02$

Figure 8. Effect of Δ of injected salt water on corridor stratified currents (center source, $\dot{m}_0 = 0.010/kg/s$).

the compartment wall was not considered until W. C. Fan and colleagues [11] made further improvement by taking into account the effect of the wall on the entrainment of a near wall plume. The following equations have resulted from this work:

$$\text{plume velocity: } U_z = (2\ \dot{Q}_0\ g/\theta C_p\rho_a T_a)^{1/3}(9\alpha/10)^{1/3}(5Z^{-1/3}/6\alpha) \tag{1}$$

$$\text{entrainment velocity: } U_b = -\alpha\ U_z \tag{2}$$

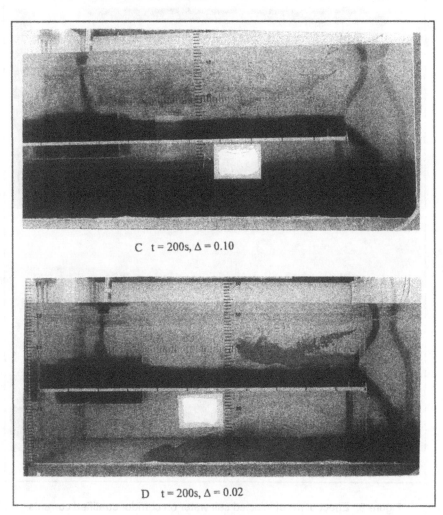

C t = 200s, Δ = 0.10

D t = 200s, Δ = 0.02

Figure 8. (Cont'd.)

point source entrainment quantity: $\dot{m}_{ep} = -\int_0^Z \theta b U_b \rho_a dz = C_0(\theta/2\pi)^{2/3}Z^{5/3}$ (3)

In the above formulas, the coefficient $C_0 = (5\alpha/6)\rho_b\pi^{2/3}(g\dot{Q}_0/C_p\rho_a T_a)^{1/3}$ $(9\alpha/10)^{1/3}$ with $b = 6\alpha\,Z/5$ the radius of the plume, α is a constant generally taken as 0.1, θ represents the effect of the wall on the entrainment quantity. If $\theta = 2\pi$,

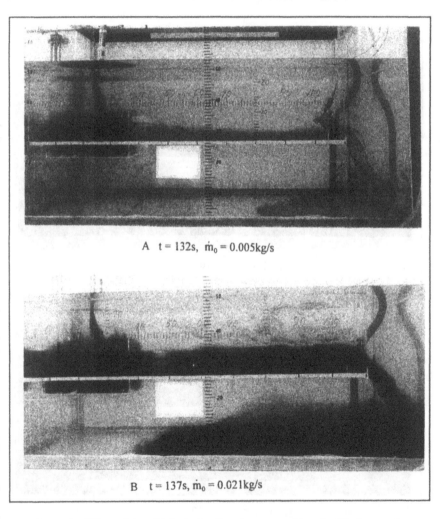

A t = 132s, $\dot{m}_0 = 0.005$kg/s

B t = 137s, $\dot{m}_0 = 0.021$kg/s

Figure 9. Effect of $\dot{m}_0$ of injected salt water on corridor stratified currents (center source, $\Delta = 0.05$).

there is no effect; if $\theta = \pi$, the fire source is on the wall surface; if $\theta = \pi/2$, the fire source is at the corner.

The non-dimensional velocity of the point source buoyant plume can be given as follows (according to the definitions of the non-dimensional variables):

$$\hat{U}_z = U_z/U_0 \tag{4}$$

$$\hat{Z} = Z/Z_0 \tag{5}$$

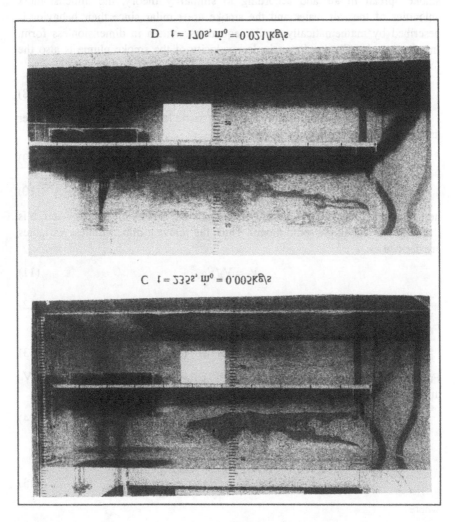

Figure 9. (Cont'd.)

where, Z_0 is the characteristic length scale, U_0 is the characteristic velocity of the smoke plume which can be written respectively as:

$$Z_0 = H, \qquad U_0 = (g\dot{Q}_0/C_p\rho_aT_aH)^{1/3} \tag{6}$$

Thus, from equations (1), (4), (5), and (6), it follows that the expression for the non-dimensional velocity of the point source buoyant plume, U_z is

$$\hat{U}_z = (2/\theta)^{1/3}(9\alpha/10)^{1/3}(5\,\hat{Z}^{-1/3}/6\alpha) \tag{7}$$

As mentioned earlier, the diffusion of salt water in fresh water is analogous to smoke spread in air and according to similarity theory, the dimensionless velocities of the salt water and the smoke are similar since their behavior is described by mathematically similar equations written in dimensionless form. In other words, (7), the dimensionless velocity of the smoke plume is also the formula of the dimensionless velocity ($\hat{V}_z$) of the salt water current,

$$\hat{V}_z = \hat{U}_z = (2/\theta)^{1/3}(9\alpha/10)^{1/3}(5\,\hat{Z}^{-1/3}/6\alpha) \tag{8}$$

The characteristic velocity and the characteristic length scale for the salt water diffusion case are defined as follows:

$$V_0 = (\dot{m}_0\,g/\rho_sh)^{1/3} \tag{9}$$

$$h_0 = h \tag{10}$$

here, m_0 is the mass flux of the salt water, ρ_s is the density of the salt water, h is the height of the single room model. Thus, the relevant dimensionless variables are defined as:

$$\hat{V}_z = V_z/V_0 \tag{11}$$

$$\hat{Z} = Z/h_0 \tag{12}$$

From equations (8), (9), (10), (11), we get

$$V_z = (2\dot{m}_0g/\theta\rho_s)^{1/3}(9\alpha/10)^{1/3}(5\,Z^{-1/3}/6\alpha) \tag{13}$$

Furthermore, from equations (2), (3), and (13), the plume entrainment velocity V_b and the entrainment mass rate $\dot{m}_e$ are given as,

$$V_b = -\alpha\,V_z, \; b = \hat{b}h_0 = 6\alpha Z/5 \tag{14}$$

$$\dot{m}_e = -\int_0^z \theta bV_b\rho_sdz = C'_0(\theta/2\pi)^{2/3}Z^{5/3} \tag{15}$$

where,

$$C'_0 = \rho_s\pi^{2/3}(\dot{m}_0\,g/\rho_s)^{1/3}(9\alpha/10)^{1/3}(6\alpha/5) \tag{16}$$

Equation (15) describes the relation between the entrainment of the salt water plume and the salt water source mass rate $(\dot{m}_0)$ as well as the corridor salt water layer (Z). Referring to the formula, the effect of the value θ on $\dot{m}_e$ is very obvious, since the value of θ is determined by the position of the salt water source. Therefore, the position of the injected salt water source is very important to the entrainment of the salt water plume into fresh water, Figure 4. Theoretically, the entrainment of the salt water plume at the center of a room should be four times that of the plume at a corner of the room. However, the real increases in the entrainment are less than the theoretical value.

Based on the analysis detailed in the previous paragraph and the experimental results, smoke velocity modeling can be achieved by assuming that the mass flux of the salt water in the corridor is

$$\dot{m} = \dot{m}_0 + \dot{m}_e \tag{17}$$

where ρ_0 is the density of the ambient fresh water, ρ_s is the density of the injected salt water, and ρ_1 is the density of the salt water current in the corridor. Thus

$$\rho_1 = (\dot{m}_0 + \dot{m}_e)\,\rho_0\rho_s/(\rho_s\,\dot{m}_e + \rho_0\,\dot{m}_0) \tag{18}$$

if the corridor width is B, then the whole volume flux per unit-width of the salt water is

$$Q = \dot{m}/\rho_1 B = (\rho_s\,\dot{m}_e + \rho_0\,\dot{m}_0)/(\rho_0\rho_s B) \tag{19}$$

Assuming that the movement of the salt water in the corridor can be described by the horizontal passage gravity current model after M. V. Chaobotov and colleagues [12]. This model suggests that when a salt water layer with volume flux per unit width Q, density $\Delta = (\rho_1 - \rho)/\rho$, thickness h_1 flows through an opening into a horizontal passageway of width B containing fresh water, the movement is initially driven by inertia force and then by a viscous force and finally by a buoyant force after the current moves for a certain distance X_t. When $X_t = 100h_1$ the general equation for the mean velocity of the salt water layer is given as

$$v/(\Delta g Q)^{1/3} = \{2+4X/[5h_1(Re_x)^{1/2}]\}^{-1/3} \tag{20}$$

if X is small or Rex is large enough, it can be assumed that $4X/[5h_1(Re_x)^{1/2}]\rightarrow 0$, then equation (20) can be simplified as

$$v = 2^{-1/3}(\Delta g Q)^{1/3} \tag{21}$$

adopting mathematical forma as given by equation (21), the expression for the mean salt water velocity can be written as

$$v = C(\Delta g Q)^{1/3} \tag{22}$$

where, C has been determined by experiment and found to have a value of C = 0.42. Assuming a gravity current which, having density ρ_1 and mass flux rate m_0, moves in a corridor containing fresh water of density ρ, then the velocity of the current in the corridor according to equation (22) is

$$v = C\{[(\rho_1 - \rho)/\rho][\dot{m}_0 g/(\rho_1 B)]\}^{1/3} \tag{23}$$

From equations (18), (19), and (22), the velocity of the salt water in the corridor is

$$v = C\{[\dot{m}_0 g/(\rho_s \dot{m}_e + \rho_0 \dot{m}_0)][(\rho_s - \rho_0)(\rho_s \dot{m}_e + \rho_0 \, \dot{m}_0)/(\rho_0 \rho_s B)]\}^{1/3}$$

$$= C\{[(\rho_s - \rho_0)/\rho_0][\dot{m}_0 \, g/(\rho_s B)]\}^{1/3} \tag{24}$$

Comparing equations (23) and (24), we can reach the conclusion that the velocity of the salt water current in the corridor induced by the room salt water source is the same as the velocity of the gravity current induced by a corridor salt water source of the same mass flux rate.

From equation (24), the dimensionless velocity of the salt water current in the corridor is

$$\hat{V} = C(\Delta/\hat{B})^{1/3} \tag{25}$$

where, $\Delta = (\rho_s - \rho_0)/\rho_0$, $\hat{B} = B/h$, $\hat{B}$ is the dimensionless width.

Also, equation (25) is assumed to represent the dimensionless velocity of the smoke in the corridor according to the scaling theory analysis discussed above. Equation (25), due to its dimensionless form, follows that the mass flux of the room source has no effect on the dimensionless salt water flow in the corridor. From the experimental results as shown in Figure 6, this conclusion is true no matter whether the point source is at the corner or in the center of the room and what the value of Δ is. Thus, assuming complete analogous behavior exists then the dimensionless velocity of the smoke in the corridor does not depend on the heat release rate of the fire source in the room. Given the characteristic velocity of the smoke (U_0) is

$$U_0 = (\dot{Q}_0 \, g/(\rho_a C_p T_0 H))^{1/3} \tag{26}$$

Then, the spread velocity of the smoke in the corridor can be written as

$$U = C\{[\dot{Q}_0 g/(\rho_a C_0 T_0 B)] \, [(T - T_0)/T_0]\}^{1/3} \tag{27}$$

where T and T_0 are temperatures of smoke layer and air layer, respectively.

Similarly, an expression for the induced air velocity into the smoke flow in the corridor can be determined by further analysis.

Suppose the salt water layer thickness in the corridor is h_1, the fresh water thickness is h_2, as shown in Figure 10, then

$$h_2 = h - h_1 \tag{28}$$

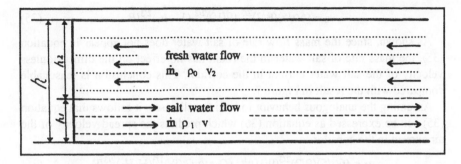

Figure 10. Movement of salt water and induced fresh water
currents in a corridor.

When the salt water current reaches the quasi-steady state, the sum of the injected salt water mass flux and the induced ambient fresh water must equal the mass flux of the salt water current in the corridor, and the mass flux of the induced fresh water equal that of the fresh water current in the corridor. The equations describing this are given as

$$\dot{m}_0 + \dot{m}_e = \rho_1 v h_1 B \tag{29}$$

and the entrainment is expressed as

$$\dot{m}_e = \rho_0 v_2 h_2 B \tag{30}$$

where, v_2 is the velocity of the fresh water.

Based on above deduction, equations (18), (24), (28), and (29), then

$$h_1 = (\rho_s \dot{m}_e + \rho_0 \dot{m}_0)/\{CB\rho_s\rho_0[(\rho_s - \rho_0)\,\dot{m}_0 g/(\rho_s\rho_0 B)]^{1/3}\} \tag{31}$$

and

$$h_2 = h - h_1 = h-(\rho_s \dot{m}_e + \rho_0 \dot{m}_0)/\{CB\rho_s\rho_0[(\rho_s - \rho_0)\,\dot{m}_0 g/(\rho_s\rho_0 B)]^{1/3}\} \tag{32}$$

Given the negligible difference between the heights of the fresh water layer surface in the room and corridor, then the induced fresh water can be calculated as follows assuming and analogous behavior to that for air equation (3),

$$\dot{m}_e = C_0' \, (\theta/2\pi)^{2/3} \, (h-h_1)^{5/3} \tag{33}$$

from equations (30) and (33), the mean velocity of the fresh water is

$$v_2 = C_0' \, (\theta/2\pi)^{2/3} \, (h-h_1)^{2/3}/\rho_0 B$$

$$= (6\alpha/5)(\theta/2)^{2/3}(9/10)^{1/3}(\dot{m}_0 g/\rho_s)^{1/3}(h-h_1)^{2/3}/B \tag{34}$$

The dimensionless velocity is

$$v_2 = (6\alpha/5)(\theta/2)^{2/3}(9/10)^{2/3}(1-\hat{h}_1)^{2/3}/\hat{B} \qquad (35)$$

As shown, since the mass flow rate of salt water does not appear in equation (35), the mass rate of salt water in the room has no effect on the dimensionless velocity of the salt water current in the corridor. This conclusion is reasonable compared with the result in Figure 7.

Assuming the analogous behavior to hold for all cases, it follows that equation (35) can be expressed as equation (36) which predicts the spread velocity of the air current in the corridor.

$$u_2 = (6\alpha/5)(\theta/2)^{2/3}(9/10)^{1/3}(\dot{Q}_0 g/\rho_a C_p T_0)^{1/3}(H-H_1)^{2/3}/B \qquad (36)$$

CONCLUSIONS

1. Salt water simulation technique has been shown to have advantage in fire safety science. The technique can be used to simulate the smoke movement within a confined space both qualitatively and quantitatively. The double-liquid-dyeing technique applied enables the visualization of the structure of the stratified salt water and fresh water layers, and permit the determination of the interface position and the measurement of the velocities of salt water and induced fresh water layers in a corridor.

2. There are two distinctive stratified flows, the origin smoke flow and the air flow in the corridor after smoke spreads from the room of fire origin. The entrainment of the smoke flow into the contra air flow is negligible while the smoke flow travels in the corridor.

3. The heat release rate and the position of the fire source together with the width of the corridor are main factors that influence the velocities of the smoke and induced air. However, the velocities of the two flows are independent of the distance moved by the flows in the corridor; in other words, these velocities are constant as the flows travel along the length direction of the corridor. Similarly, the thicknesses of the smoke and air layers are the same along the entire corridor.

4. The non-dimensional velocities of the smoke and induced air flows are independent of the heat release rate of the fire source in the room.

5. Based on the experimental results and scaling theory, and by the application of the single room point-source plume model and the passage gravity current model, velocity equations concerning the movement and velocity of the smoke and the induced air have been suggested.

NOMENCLATURE

B   Reduced-scale corridor width
b   Point plume radius
C_p Specific heat

Fr	Froude number
G	Ratio of length scales cubed
g	Gravitational acceleration
H	Full-scale corridor height
h	Reduced-scale corridor height
h_1	Salt water layer thickness in corridor
h_2	Induced fresh water layer thickness in corridor
L	Diameter of heat source
l	Diameter of salt water source
$\dot{m}_e$	Corridor entrainment mass flux rate
$\dot{m}_0$	Mass release rate of salt source
$\hat{\dot{m}}_0$	Dimensionless mass rate of injected salt water
p	Pressure
p_0	Ambient pressure
Pr	Prandtl number for air
Q	Salt water volume flux per unit-width in corridor
$\dot{Q}_0$	Heat release rate
$\hat{\dot{Q}}_0$	Dimensionless heat release rate
Re	Reynolds number
Sc	Schmidt number for salt water in fresh water
T	Smoke temperature
T_0	Ambient air temperature
t	Time
$\hat{t}$	$t\dfrac{V_0}{h}$
U_0	Characteristic velocity of smoke in air
U_z	Point source smoke plume velocity
$\hat{U}_z$	Dimensionless point source smoke plume velocity
U_b	Point source plume entrainment velocity
u	Velocity scale
v	Corridor salt water current velocity
$\hat{v}$	Dimensionless corridor salt water current velocity
v_2	Corridor fresh water current velocity
$\hat{v}_2$	Dimensionless corridor fresh water current velocity
V_0	Characteristic velocity of salt water in fresh water
V_z	Point source salt water plume velocity

$\hat{V_z}$ Dimensionless point source salt water plume velocity

$\hat{x}$ $\dfrac{x}{h}$

ρ Density

ρ_0 Ambient fresh water density

ρ_a Ambient air density

ρ_s Salt water density

ρ_1 Corridor mixed salt water current density

Δ Normalized density difference $(\rho - \rho_0)/\rho_0$

REFERENCES

1. K. Towage, *Fire Behaviour in Rooms*, BRI Report No. 27, Building Research Institute, Tokyo, 1958.
2. J. G. Quintiere, Some Observations on Building Corridor Fires, *Fifteenth Symposium (International) on Combustion*, The Combustion Institute, 1975.
3. B. J. McCaffrey and J. G. Quintiere, Fire-Induced Corridor Flow in a Scale Model Study, *Symposium on the Control of Smoke Movement in Building Fires, Proceedings*, BRE, Garston, United Kingdom, 1975.
4. J. G. Quintiere, B. J. McCaffrey, and W. Rinkinen, Visualization of Room Fire Induced Smoke Movement and Flow in a Corridor, *Fire and Materials*, 2, p. 18, 1978.
5. H. R. Baum, J. G. Quintiere, and K. D. Stickler, *Salt Water Modeling of Fire Induced Flows in Multicompartment Enclosures*, NBSIR 86-3327, U.S. Dept. of Commerce, National Bureau of Standards, National Engineering Laboratory, Centre for Fire Research, Maryland, March 1986.
6. B. K. Ma and R. J. Zhang, Criteria of Salt Water Simulation in a Confined Space, *Fire Safety Science Journal*, 3:1, pp. 52-55, 1994. (in Chinese)
7. H. P. Zhang, R. J. Zhang, B. K. Ma, W. C. Fan, and R. Huo, Experimental Study of Salt Water Simulation of Smoke Movement in a Confined Space, *Fire Safety Science Journal*, 3:2, pp. 48-57, 1994. (in Chinese)
8. E. E. Zukoski, T. Kubota, and B. Celeyen, Entrainment of Fire Plumes, *Fire Safety Journal*, 3, 1980.
9. B. R. Morton, G. Taylor, and J. S. Turnet, *Turbulent Gravitational Convection from Maintained and Instantaneous Source*, Proc. of Rord Soc., A234, 1, London, 1956.
10. W. D. Baines and J. S. Turnet, *Turbulent Buoyant Convection from a Source in a Confined Region*,
11. W. C. Fan, Q. A. Wang, R. J. Zhang, and R. Huo, *Introduction to Fire Safety Science*, Hubei Press of Science and Technology, 1993. (in Chinese)
12. M. V. Chobotov, E. E. Zukoski, and T. Kubota, *Gravity Current with Heat Transfer Effects*, NBS-GRC-87-522, California Institute of Technology, Pasadena, California, December 1986.

Contributors

ALI, FARIS A., graduated with a first class Civil Engineering degree at the University of Technology, Iraq, in 1983. After three years of working as a structural engineer in Iraq where he was involved in the structural design of several projects in Baghdad, he started his postgraduate study of structural engineering in the former Soviet Union obtaining his MSc in 1988 and Ph.D. in Moscow, 1992. In the same year he arrived in the United Kingdom and worked on the performance of steel reinforced concrete structures in fire for nine years. Initially he worked as a visiting lecturer at City University (London), and currently he works as a Research Officer at the Fire Research Center, University of Ulster, UK. Research interests are: creep of concrete, thin-walled steel structures, behavior of steel columns in fire, spalling of high strength concrete under high temperatures, and performance of axially and rotationally restrained steel frames in fire.

BEARD, ALAN, studied physics at Leicester University and in 1972 was awarded a Ph.D. in theoretical physics from Durham University for his thesis entitled *A calculation of Neutron-Deutron scattering using the SU(3) basis of three particle states*. On completion of his Ph.D., he conducted post-doctoral research in medical physics at Exeter University and the University of Wales, Dental School, Cardiff. In 1977 he started research in the Fire Safety Engineering Unit at Edinburgh University, leaving in 1995 to become a lecturer at Heriot-Watt University. He has acquired a wide knowledge of fire modeling and has conducted work for both government departments and industrial companies. He was instrumental in the establishment of a Home Office working group, the Fire Models Context Group, of which he is a member. The Group is concerned with the setting up of standards and a regulatory framework for the use of fire models. His papers have been used as key references by the International Standards Organization and some of his work has been translated into Japanese. He is also a member of the Fire Study Group of the Institution of Structural Engineers. For over 20 years Dr. Beard's research has centered around fire modeling. He has constructed both probabilistic and deterministic models of fire development. His previous work includes a six-year project on the probabilistic modeling of fire growth in hospital wards, funded by the Department of Health and the fore-runner of the Engineering and Physical Sciences Research Council (EPSRC).

177

BIRK, A. M., is a Professor in the Department of Mechanical Engineering at Queen's University in Kingston, Ontario, Canada. He received his Ph.D. degree from Queen's University in 1983 and joined the faculty there in 1986. Before joining Queen's faculty he spent five years working in the oil and gas, transportation research, and engineering consulting fields. He is a registered Professional Engineer in the Province of Ontario. Professor Birk's research interests include modeling of the thermal response of pressure vessels exposed to fire, field fire testing of pressure vessels, testing and analysis of thermal protection systems, and testing and analysis of BLEVEs and their hazards. Dr. Birk has published widely in these areas and has served as consultant for various public and private organizations.

CLARK P. J., is at the Heriot-Watt University, Edinburgh U.K.

CORBETT, GLENN P., currently serves as an Assistant Professor and Coordinator of Fire Science at John Jay College of Criminal Justice. Additionally, he serves as a Captain in the Waldwick, N.J. Fire Department, Vice President of the New Jersey Society of Fire Service Instructors, Technical Editor of *Fire Engineering* magazine, and a member of the New Jersey State Fire Code Council. In 1999, he was appointed to the "America Burning: Re-commissioned" panel by FEMA Director James Lee Witt. Corbett previously served as the Administrator of Engineering Services of the San Antonio, Texas Fire Department.

FAN, W. C., is at the State Key Laboratory of Fire Science, China.

FERNANDEZ, ABEL A., is Associate Professor and Director of the Engineering Management Program at the University of the Pacific, Stockton, California. He holds the Ph.D. degree in Industrial Engineering from the University of Central Florida, M.E. and B.S. degrees in Electric Power Engineering from Rensselaer Polytechnic Institute (RPI) and an M.B.A. also from RPI. He has over 12 years of system engineering and project management experience with TRW, Inc. and the Harris Corporation. His final position at Harris was Director of Product Marketing, in which he was responsible for applications engineering and strategic planning for the Controls Division of the Harris Corporation. He obtained his Ph.D. degree under sponsorship of a NASA Graduate Student Research Fellowship, working at Kennedy Space Center. His general field of interest is the analysis of project management problems in which uncertainty should be considered. Specifically, these include project risk analysis, stochastic project scheduling, modeling of decision analysis problems in project management contexts, and allied problems within project management. He is a member of the IIE, Decision Sciences Institute and the American Society of Engineering Management, and is a registered Professional Engineer in the State of Texas.

FORSYTHE, THOMAS J., PE is a graduate from the Illinois Institute of Technology with undergraduate training in Fire Protection and Safety Engineering. After years in the fire protection contracting arena, Mr. Forsythe entered the consulting field, and is currently Principal-In-Charge and Managing Engineer of the Gage-Babcock & Associates, Inc. San Francisco area office. Mr. Forsythe has long participated in GBA's work on the fire protection issues related to alternative fuel vehicle maintenance and operation, and sits on several NFPA Technical Committees, including the TC on Vehicular Alternative Fuel Systems (NFPA 52/57), the TC on Automotive and Marine Service Stations (NFPA 30A), the TC on Parking Structures (NFPA 88A), the TC on Hanging and Bracing Water Based Suppression Systems (NFPA 13) and the TC on Water Mist Systems (NFPA 750). Mr. Forsythe has written numerous papers and articles, and has spoken to national organizations in the fire protection community. He is a registered Fire Protection Engineer in the State of California.

FOX, HERBERT, earned a B.S. degree in Aeronautics and Astronautics from the Massachusetts Institute of Technology in 1960, and M.S. and Ph.D. from Polytechnic Institute of Brooklyn in 1962 and 1964 respectively. He served as Chairman of the Department of Mechanical Engineering and was the first Dean of Science and Technology at the New York Institute of Technology. Later he joined Pope, Evans and Robbins, Inc., where he served as Vice-President of Consulting Services. He returned to academic life at New York Institute of Technology as Senior Vice-President for Scientific Research and in 1992 again returned to research and teaching as Professor of Mechanical Engineering. Research projects have included seeking alternative fuels for the Metropolitan Transportation Authority Long Island Bus Division and the Norwalk Transit District including consideration of accidental and explosive ignition. Dr. Fox has published over 50 articles and two books.

JACOBS, DERYA A., is Associate Dean of College of Engineering and Technology at Old Dominion University, Norfolk, Virginia. She has a B.S. in Computer Science, M.S. and Ph.D. in Engineering Management from University of Missouri-Rolla. Dr. Jacobs served as the Graduate Program Director for Engineering Management, Interim Associate Dean for the College of Engineering and Technology, and as the Acting Chair in the Department of Engineering Management. Prior to joining Old Dominion in 1990, Dr. Jacobs worked as a Research Engineer at Missouri Enterprise and the Center for Technology Transfer and Economic Development in Rolla, Missouri. Dr. Jacobs' research interests include operations research, applied artificial intelligence and neural networks. She has worked on projects sponsored by various industry and government agencies including NASA Langley Research Center, Lockheed Martin Vought Systems, Thomas Jefferson National Laboratory, Naval Undersea Warfare Center Detachment, Department of Army, Joint Training, Analysis, and Simulation

Center, and American Red Cross. She is currently President-Elect for the American Society for Engineering Management.

KASHIWAGI, TAKASHI, is at the National Institute of Standards and Technology, Center for Fire Reaearch, Gaithersburg, Maryland.

KAUFFMANN, PAUL J., is an Assistant Professor in the Department of Engineering Management at Old Dominion University. Prior to his academic career, he worked in industry where he held positions as Plant Manager and Engineering Director. Dr. Kauffmann received a BS degree in Electrical Engineering and MENG in Mechanical Engineering from Virginia Tech. He received his Ph.D. in Industrial Engineering from Penn State and is a registered Professional Engineer. His research interests include cost analysis, technology planning, and operational improvement.

KEATING, CHARLES B., is an Assistant Professor of Engineering Management at Old Dominion University. He has a B.S. in Engineering from West Point, an MA in Management and Supervision University. His industrial experience includes 12 years of management and supervisory positions. He served as a military officer for over five years in numerous staff and command positions. Most recently his research has focused on Systems Engineering & Methodologies, Analysis & Design of Organizational Knowledge Systems, and Project Management Systems.

KERWIN, RALPH, is a Loss Prevention Representative with Industrial Risk Insurers, a Division of GE. He holds a Masters Degree in Mechanical Engineering and has 13 years of experience with five protection and loss prevention engineering activities in construction, consulting, municipal and insurance contexts. Mr. Kerwin lives in the San Francisco Bay area.

O'CONNOR, DAVID, is a Senior Lecturer in Structures at the University of Ulster. Current research projects include the behavior of structural elements and assemblies under combined loading and the elevated temperatures associated with fires—brickwork compartment walls, concrete floor slabs, structural steel assemblies, and the spalling of concrete in fires. He also has a research interest in fire severity in natural fires and standard fire tests, the measurement of heat flux in fires and fire engineering design methodologies.

SANTOS-REYES, J., is a Ph.D. student in the Department of Civil and Offshore Engineering at Heriot-Watt University, Edinburgh, Scotland. He has been involved in the offshore industry for two years. He obtained a MSc in Thermal Power and Fluid Engineering at the Mechanical Engineering Department at UMIST, England in 1996.

SFORZA, PASQUALE M., received his Ph.D. in Aeronautics from the Poly-technic Institute of Brooklyn in 1965. He served on the faculty of Polytechnic from 1965 to 1998, was appointed to professorship of Mechanical and Aeronautical Engineering in 1977 and served as Head of Department from 1983 to 1986 and as Head of the Aerospace Department from 1988 to 1995. Dr. Sforza is a member of the New York Academy of Science. He was written over 100 papers and reports and holds three patents in the areas of theoretical and experimental fluid mechanics, turbulence, energy and heat transfer. He has also served on the Advisory Committee to the New York State Legislative Commission on Science and Technology. He has received the 1977 Technology Achievement Award from AIAA and an outstanding Technical Paper Award at the 1992 Joint Inter-society Energy Conversion Engineering Conference.

SHIELDS, T. JIM, is a graduate of Edinburgh University and the University of Ulster and is a Professor of Fire Safety Engineering at University of Ulster. He is Director of the Fire SERT Centre and has been involved in teaching and conducting research in Fire Safety Engineering for over twenty-five years. He is coordinator of Unit of Assessment 33 Built Environment which achieved a rating of 4 in the 1992 and 1996 research assessment exercises. His current research interests are extensive including the behavior of materials, components, and people in fire. With colleagues at Fire SERT he organized and hosted the first international symposium on human behavior in fire in September 1998. He was awarded the Association of Building Engineers "Man of the Year" award in 1995 for his work in fire safety engineering. On the international scene he has contributed to the work of the ISO the Conseil du Batiment CW14 Fire Safety Engineering, World Health Organisation and the International Association for Fire Safety Science. He was a Visitor to BRE Fire Research Station, and has contributed to the work of BSI committees in Fire Safety Engineering. Jim was appointed as a member of the Fire Authority for N.I. which is responsible for the provision of fire services in Northern Ireland.

SILCOCK, GORDON, W. H., is a Senior Lecturer in Fire Safety Engineering. He is a chartered Physicist and graduate of the Queens University of Belfast where he studied physics and mathematics. Having been involved initially in education he has for the last eighteen years been actively involved in Fire Safety Engineering Research and Education at University of Ulster.

SIZEMORE, DENNIS C., is at the City of Hampton VA Fire Rescue Division, Virginia.

SMITH, RICHARD L., is currently retired. His previous areas of research include developing a fundamental risk assessment methodology for the FAA based upon

Artificial Intelligence techniques of dealing with knowledge that has a large amount of uncertainty associated with it. This work used influence diagrams and Bayesian Probability Theory. He developed the expert system EXPOSURE that produces solutions for the problem of preventing the spread of fire between buildings. This program showed that existing fire codes had potentially serious shortcomings, sometimes calling for unnecessarily expensive precautions and sometimes resulting in unsafe conditions. He also developed the concept of the Calculus of Fire Safety. This calculus would be a fifth-generation language that would allow users of various levels of knowledge of fire technology to input information and obtain answers to fire safety questions using concepts they normally use. Dr. Smith was a member of the IEEE and OSA (Optical Society of America). He is a former vice president of the ART users' group and member of the Board of Directors of the Laser Institute of America. Dr. Smith is the recipient of several honors and awards including the 1987 Harry C. Bigglestone Award for Excellence in Written Communication of Fire Protection Concepts, the Department of Commerce Bronze Medal Award for Superior Federal Service, and a Commerce Department Fellowship. He received his Ph.D. in Physics from the University of Maryland, his M.S. in Physics from the University of Illinois, and his B.S. in Physics from the University of Kansas.

ZHANG H. P., is at the State Key Laboratory of Fire Science, China.

ABOUT THE EDITOR

Paul R. DeCicco is Professor Emeritus at the Polytechnic University (New York) where he taught courses in Civil and Fire Protection Engineering and has served as Director of the Center for Fire Research. As principal Investigator in a wide range of fire research projects he has directed full scale fire tests for the New York Fire and Building Departments in connection with the development of building and fire codes. He has also been engaged in mathematical and physical modeling of fire phenomena in various building occupancies, and in the study of fires in large spaces. He is a fellow of the American Society of Civil Engineers and the Society of Fire Protection Engineers and currently serves as Executive Editor of the *Journal of Applied Fire Science*. Professor DeCicco is a registered Professional Engineer and has practiced Civil and Fire Protection Engineering for over forty years. He is co-author of *Making Buildings Safer for People* (Van Nostrand, 1990), and has published a number of papers on fire protection engineering. He has been honored for his research work in fire protection of high-rise buildings and in the improvement of fire safety in high risk urban residential buildings. He is currently engaged as a consultant in the investigation of fires and occasionally serves as an expert witness in fire litigations.

Index